T0244430

"Charles Lachman weaves the incredible story of the capture by American forces of a German U-boat and its secrets during World War II, an operation which allowed Allied forces to shock the German Navy. Richly detailed with undeniable suspense and action, *Codename Nemo* is destined for the nonfiction bestseller lists."

—Bill O'Reilly,
#1 *New York Times* bestselling author of the Killing Series

"Crisp as a torpedo striking the water, *Codename Nemo* pulls you along with a deeply personal account of the hunters on both sides of an amazing drama."

—Walter R. Borneman,
author of *The Admirals: Nimitz, Halsey,
Leahy, and King—The Five-Star Admirals Who Won the War at Sea*

"A relentless, pressure-packed plunge into the depths of war. *Codename Nemo* is a storytelling tour de force—indeed, the quintessential story of the Battle of the Atlantic, rendered in taut prose, and with an immediacy and intimacy that all but makes a participant of the reader. In the wake of this story, you'll feel a profound sense of gratitude to the men who went after *U-505*, and to Charles Lachman for bringing them back."

—James Sullivan,
author of *Unsinkable: Five Men
and the Indomitable Run of the USS* Plunkett

"The best missions involving submarines often start with an outlandish idea, and the very best make a hell of a story. *Codename Nemo* does both."

—Sherry Sontag,
coauthor the *New York Times* bestseller
Blind Man's Bluff: The Untold Story Of American Submarine Espionage

"*Codename Nemo* is a pulse-pounding tale of high-stakes espionage and daring courage, detailing the pursuit and capture of a German U-boat at the height of World War II. Charles Lachman masterfully builds a cast of characters, German and American, whose destinies intersect in the perilous waters of the Atlantic. The vivid descriptions of life aboard a U-boat immerse you in the claustrophobic, terrifying world of underwater warfare. As the tension builds, with each 'ping' of the Sonar, the thrilling plot keeps you turning the pages. A riveting narrative combining historical research with visceral scenes, *Codename Nemo* is a must-read for anyone in search of a thrilling maritime adventure."

—Andrew Dubbins,
author of *Into Enemy Waters: A World War II
Story of the Demolition Divers Who Became the Navy SEALs*

"What a terrific read from Charles Lachman! A suspenseful, fast-paced but little-known saga of hide-and-seek between a US 'baby flattop' and a German U-boat during World War II. Who can resist an Irish-American commander nicknamed 'Full Flaps' by his crew—always beyond his hearing, of course—because of his protruding ears, who embarks with his dog, a border collie named Flairby, instigates nail-biting nighttime take-offs and landings on his carrier for the first time in Naval history . . . and if that's not enough, then proposes some cockamamie scheme of commandeering a German sub filled with secret stuff by actually attempting to board her with a nine-man team! It's a wild, engrossing ride from start to finish with extraordinary details and insights into daily life—clashes, arguments, even suicide—aboard both German U-boats and American ships during the Battle of the Atlantic. This one is a winner!"

—Carole Engle Avriett,
author of *Coffin Corner Boys: One Bomber,
Ten Men, and Their Harrowing Escape from Nazi-Occupied France*

Also by Charles Lachman

Footsteps in the Snow:
One Shocking Crime. Two Shattered Families.
And the Coldest Case in US History.

A Secret Life: The Lies and Scandals
of President Grover Cleveland

The Last Lincolns: The Rise & Fall
of a Great American Family

In the Name of the Law

THE HUNT
FOR A
NAZI U-BOAT
AND THE ELUSIVE
ENIGMA MACHINE

CODENAME NEMO

CHARLES LACHMAN

DIVERSION
BOOKS

NEW YORK

Diversion Books
A division of Diversion Publishing Corp.
www.diversionbooks.com

Diversion Books and colophon are registered
trademarks of Diversion Publishing Corp.

For more information, email info@diversionbooks.com

First Diversion Books Edition: June 2024
Hardcover ISBN 978-1-635-76871-8
e-ISBN 978-1-635-76867-1

Book design by Aubrey Khan, Neuwirth & Associates, Inc.

Printed in the United States of America
10 9 8 7 6 5 4 3

Diversion books are available at special discounts for bulk purchases
in the US by corporations, institutions, and other organizations.
For more information, please contact admin@diversionbooks.com.

The publisher does not have any control over and does not assume any
responsibility for author or third-party websites or their content.

To Arthur I. Lachman-Heath
and my wife, Nancy Glass

The only thing that really frightened me
during the war was the U-boat peril.

—SIR WINSTON CHURCHILL

CONTENTS

AUTHOR'S NOTE

All the dialogue in *Codename Nemo* is taken verbatim from oral histories, autobiographies, interviews, diaries, letters, contemporary newspaper quotations, and US government tape recordings of World War II battles. Nothing has been invented.

KEY FIGURES

THE BOARDING PARTY

Lieutenant Junior Grade Albert David (Mustang)

Signalman Second Class Gordon Hohne (Flags)

Chief Motor Machinist's Mate George Jacobson (Chief)

Torpedoman Third Class Arthur Knispel (Tommy Gun)

Motor Machinist First Class Zenon Lukosius (Zeke)

Gunner's Mate First Class Chester Mocarski (Chesty)

Boatswain's Mate Second Class Wayne Pickels (Mack)

Electrician's Mate Third Class William Riendeau (Rhode Island)

Radioman Third Class Stanley Wdowiak (Stash)

 And,

 Master-at-Arms Leon Bednarczyk (Ski)

THE BRASS

Captain Daniel V. Gallery (Cap'n), USS *Guadalcanal*

Commander Earl Trosino (Earl), chief engineering officer,
USS *Guadalcanal*

Commander Jessie Johnson (XO), executive officer,
USS *Guadalcanal*

Dr. Henry Morat (Doc), medical officer, USS *Guadalcanal*

Commander Dudley Knox (Bangkok), USS *Chatelain*

Commander Kenneth Knowles (Soothsayer),
anti-submarine warfare

Rear Admiral Francis Low (Frog), chief of staff

Fleet Admiral Ernest King (Uncle Ernie), COMINCH,
chief of Naval Operations

THE GERMANS

Kapitänleutnant Axel-Olaf Loewe (the Lion)

Kapitänleutnant Peter Zschech (Skipper)

Kapitänleutnant Harald Lange (the Old Man)

Machinist Second Class Hans Goebeler (Hans)

U-505 Cook Anton Kern (Toni)

Óberleutnant Josef Hauser (Raccoon)

Radio operator Gottfried Fischer (Gogi)

U-515 Kapitänleutnant Werner Henke (Lone Wolf)

Grand Admiral Karl Doenitz (Onkel Karl; a.k.a. Uncle Karl)

And,

Junior Seaman Ewald Felix (the Defector) of Poland

Eric Munday (Sapper) of England

Jeanette (the Girlfriend) of France

PROLOGUE

A German torpedo is slicing through the water, seconds away from blowing up the US cargo ship *Thomas McKean*.

It's Monday, 7:23 a.m., just after sunrise.

Thomas McKean is on a zigzag course, intended by the skipper to make her a moving target, harder to hit. Right now, she is four hundred miles northeast of Puerto Rico, on her maiden voyage. There are sixty-two men on board, and the ship's hull is loaded with war supplies—tanks, trucks, food, and more. On the top deck are eleven Douglas medium bombers. Ammo is stored in hold #1. Aviation fuel fills hold #2. She is a ticking time bomb.

There are six lookouts posted on deck—two forward, two aft, and two on the bridge.

But none of them see it coming.

Wham!

Chief Engineer Thomas McCarthy has just lathered his face for a shave, and the impact nearly knocks him over. McCarthy has served on ships for thirty years, and he knows the situation. Hell, it happened to him just six weeks earlier when another freighter he was aboard was torpedoed. Chrissakes, not again.

General quarters ring out and all hands report to their battle stations. McCarthy hightails it to the engine room to check on his three morning-watch crewmen—the first assistant engineer, the oiler, and the water tender who takes care of the boilers. McCarthy looks around. They're all gone. As he was going down, they must have been going out. Good for them.

McCarthy hurries up the ladder and finds Master Captain Mellin Respess on the bridge, absorbing all the damage reports coming at him at a breakneck clip. This is nothing new for Respess, either. Three months ago, Respess was skipper of a freighter that was torpedoed off Cape Hatteras, North Carolina—he, too, knows the destructive power of these weapons.

"How hard are we hit?" McCarthy asks. "It looks to me like the whole stern is gone."

"Climb over and see what you can see," Respess commands.

McCarthy runs down to the deck, hops over the side of the ship, and starts climbing down the cargo netting. The netting is an innovation. In the old days, you used a ladder to get to the lifeboats, meaning one sailor at a time. Pretty slow going in an emergency. Now, with netting, twenty men can scale down together straight into the lifeboats.

"What does it look like?" the skipper hollers over the side of the ship.

McCarthy turns his head this way and that to get a clear look at the damage. Oh, boy. The torpedo slammed into the ship about fifteen feet below the surface, leaving a gaping hole. Now the hull is wide open in the shape of a perfect "V." From what McCarthy can tell, the propeller isn't there anymore.

"It looks like she's finished," McCarthy shouts back at the captain. Yep, the ship is a goner.

McCarthy clambers up the netting. There are things he must take care of right away. First, he returns to the engine room and shuts down the engine. He orders the water tender, a sailor named Short, to close the fuel pump. Then, McCarthy shuts off the throttle valve.

When he's done, he returns to the captain on the bridge and makes it official. "Well, we are a dead pigeon."

The order goes out to abandon ship. Respess quickly collects all the sensitive documents that he can gather into his arms to keep them from falling into enemy hands should the Nazis board the ship.

"What about the distress message?" McCarthy asks.

"The message has gone out that we got hit," Respess replies.

The SOS was sent by Sparks, the radio operator. By now, the US Navy is aware that the freighter *Thomas McKean* has been hit by a torpedo and is about to go down at 22° 10'N latitude and 60° 00'W longitude.

McCarthy gives a brisk nod. As he turns to go, the captain tells him to hang on a second. He's got one more order.

"Grab a case of cigarettes."

McCarthy and the captain head over to Lifeboat Station #1 and McCarthy tosses over the case of cigarettes. He and the skipper climb down the netting and join the other men who are waiting for them in the lifeboat. McCarthy is pissed. He left everything behind on the *Thomas McKean*—even his sneakers, which are still on the bathroom floor. All he has is what he's wearing on his back.

"Is everyone accounted for?" the captain asks.

"Yes," McCarthy says.

"Pull away and we can check on those other boats."

Just as Lifeboat #1 starts rowing away, they hear a cry coming from the deck of *Thomas McKean*.

It's Sparks, the radio operator. What the hell? Sparks kept transmitting those SOS signals even after the order went out to abandon ship. Good man, that Sparks, but now he's screwed. McCarthy and the others yell at Sparks to jump. The ship is tilting and sinking, and the deck is getting so steep that Sparks has to get on his hands and knees for balance.

"Jump!"

"Jump!"

Captain Respess cuts loose the lifeboat's painter rope so Sparks will have something to grab. Finally, Sparks makes the leap overboard, splashing down hard. Lifeboat #1 comes around, and they fish the radio operator out of the water. Respess orders the crew to look for any other survivors but there's nobody in the water.

Just then, about a mile off, a submarine breaks the surface.

It's a German U-boat.

She's about a mile from the crippled *Thomas McKean*. All at once, she opens up with three-inch and four-inch guns. It's time to finish off this American freighter.

The first four shells fall short. Lifeboat #2 is in the direct line of fire, and one shell comes way too close. They start rowing for dear life. The U-boat moves in, firing the entire time. Now the Germans can't miss. They direct their gun power on the hold where the dynamite and

.50-caliber antiaircraft ammunition are stored. They next focus their aim on the hold containing the combustible aviation fuel.

Flames and smoke engulf *Thomas McKean*. The U-boat crew seems to be having fun toying with the Americans. They unload a barrage of seventy-two rounds, spaced two to fifteen seconds apart.

Finally, *Thomas McKean* goes down, sinking twelve thousand feet into oblivion.

Then, and only then, does the U-boat cease fire.

Roland Foster Jr. is second mate on *Thomas McKean*, which makes him third in command. He's the ship's navigator and sometimes, when he has to be, he's the medical officer. From his position in Lifeboat #2, Foster can make out the U-boat commander standing on the conning tower. He's got a megaphone in his hand.

"Please come alongside," the German calls out in surprisingly decent English. Foster wonders if he heard right. Did this German really say "please" just after bombing his ship to kingdom come?

Then the U-boat commander hands the megaphone to a junior officer—an interpreter—who speaks even better English. He's wearing just shorts, no shirt, no insignia whatsoever. He's got a big, bushy beard, like a pirate who has been at sea way too long.

Through the megaphone, the German asks: "American ship?"

"Yes," Foster yells back.

"What kind of American ship?"

"An American merchantman."

"*Gut.* Where are you bound for?"

Before Foster can say anything more, one of the guys on the lifeboat pipes up: "Trinidad!" Technically, that is accurate. Trinidad for refueling, but the ultimate destination is delivering war supplies to the Soviet Union, which is better left unsaid.

"You carrying munitions—yes?" The glower from the interpreter lets Foster know he'd better not bullshit his way out of this; it's obvious the *McKean* was packed with ammo.

Foster starts taking mental notes so he can relay to Navy intelligence everything he observes about the enemy vessel. For one thing, the U-boat is about two hundred feet long yet impressively maneuverable. She's carrying a swastika, but there's also an unusual touch: a lion painted on the side of the conning tower. The lion is standing on his hind foot, and in his front paw holds a hammer. The tail holds a torch. The U-boat seems freshly painted.

And the Germans are a stunning sight—suntanned and in good physical condition. Where have they been, on vacation?

The better question is . . . What's next? Will the Nazis open fire on them with machine guns? Foster has to wonder. It could happen. Anything could happen.

The Germans see there's a seriously injured man in Lifeboat #2: Russell Funk. They toss over liquid sleeping medication and some bandages wrapped in wax paper. As long as the Germans are in an amiable mood, Foster figures it doesn't hurt to ask the correct course to the nearest body of land.

The interpreter shakes his head. "No, we cannot do that. We haven't got the time."

That sure got lost in translation. Did he think Foster was asking for a lift, or a tow?

"I didn't say that," Foster shouts back. "I said the nearest course to land."

The interpreter nods. Now he understands. All he offers is an obvious suggestion. "Steer *mit* the wind." Speaking in German, the interpreter turns to his commander and mutters, *Sie warden gefunden*—"They will be found."

With that, the U-boat departs.

Captain Respess orders Lifeboat #1 to pull up to Lifeboat #2 to check on everyone.

Russell Funk isn't breathing. Foster says, "I think I got a dead man here, Captain."

Respess tells McCarthy, his most seasoned hand, "Chief, you'd better look at him."

McCarthy knows there's one sure way to find out if the wiper is dead. "Did you burn his feet with a cigarette butt?" McCarthy asks Foster.

"No."

"All right, let's try that."

They do so. No reaction.

Then they burn Funk's fingertips. Still nothing. Not even a flinch. No sign of life whatsoever. Even his eyes are dead.

They find a few hooks and secure them to Funk's ankles and arms and gently lower him into the sea for a fitting burial.

In total, four crew members are dead. That makes fifty-eight men on four lifeboats who need to find a way back to the United States. There's no time for mourning or second-guessing. Respess makes sure all the lifeboats are equipped with the gear necessary for a long ordeal at sea: charts, sextant, water, food, medical equipment, and sails. And, yes, cigarettes, which they divvy up. The only lifeboat outfitted with a motor and a full tank of gas is Foster's Lifeboat #2, but when they try to start the motor, all it does is sputter. Just the other day the motor was tested and worked fine. Fuck. Now they'll have to rig a sail. McCarthy mutters something about how nobody knows what the hell they're doing anymore and hops from his lifeboat to #2. He turns on the ignition, gets gas into the carburetor, spins the engine, and, like magic, the motor roars to life. The chief sure knows engines.

Respess checks his chart on the bobbing lifeboat.

"Here we are," he says, pointing to their current position. "We have to make it southerly as much as we can to land in the islands." Just their luck, the wind is swinging westerly. The closest land is Hat Island, also known as Sombrero Island, about two hundred miles away. It's the northernmost island of the Lesser Antilles, inhabited by dwarf geckos, lizards, and not a single human being. About twenty-four miles from Hat Island is a dot in the ocean called Dog Island, inhabited by feral goats and not much else.

The captain instructs everyone to stay within sight of one another. Since Foster's is the only motorized boat, Respess orders him to consolidate all the boats at sunset, should any of them lag behind.

Off they go, due south—God willing, to the Antilles.

And out there in the depths of the Caribbean is the U-boat that blew up their ship, heading west to find her next prey.

Her name is *U-505*.

CODENAME NEMO

ONE

"WHERE THE HELL IS FREETOWN?"

Fourteen Months Earlier

When Kapitänleutnant Axel-Olaf Loewe sees *U-505*, it is love at first sight. With her coat of light gray paint and dark gray trim, there is not a mark on the boat. The diesel engine sparkles, the brass fittings gleam, the batteries are fresh from the factory.

Loewe comes from a seafaring family. His father, also named Axel-Olaf, was a naval officer in the German Imperial Navy during World War I. In this new world war, Loewe's brother was an officer on a U-boat in the English Channel—until his capture, which landed him in a prisoner of war camp in Canada.

Loewe has great regard for the sea and natural instincts for maneuvering big ships. He graduated from the German naval academy in 1928, ranking second in a class of thirty-nine cadets. He served on the cruisers *Emden* and *Konigsberg*, which, before the war, took him to San Diego and Los Angeles on missions to demonstrate the growing might of the German Navy. He has toured Hollywood and speaks English. He and his wife, Helga, have a solid marriage. "A very brave seaman's wife," Loewe calls her. They have a daughter, Karin, born in 1939, and a newborn son named Axel.

Loewe served as executive officer on *U-74*. Now, at age thirty-two, he is given his first command—*U-505*. He reports for duty at the Deutsche Werft Shipyard in Hamburg.

U-505 is a large submarine, carrying twenty-two torpedoes, known as "fish" by some foreign navies and as "eels" by the Germans. Fifty men are on board. *U-505* can go more than eight thousand miles without refueling. She looks like a steel cigar, and she flies the Nazi War Eagle clutching a swastika, the flag of the Kriegsmarine, the new name of the German Imperial Navy (from 1935 to 1945).

It is August 26, 1941.

On this day, *U-505* is formally commissioned. Loewe orders his coat of arms—a rampaging lion—painted on the submarine's tower. The beast's front claw swings an axe and the tail carries a torch. The axe is the insignia of Loewe's cadet class. The big cat symbolizes Loewe's name, which in German means "lion"—amusing considering Loewe is humble and even-tempered. He is of medium height with thick, unruly hair under his captain's cap and a genuine smile. He is well regarded for steady leadership. And if he ever questioned serving a monstrous regime like the Nazis, he kept it to himself.

Loewe stands proudly atop the tower to address his crew.

"Comrades, as commandant of *U-505*, I have come here to Hamburg in order, with your help, to take a boat to the front after our short shakedown and war training maneuvers. It will be a hard life—have no illusions about that—but with a well-disciplined crew we will have our successes."

That night the crew parties. The next morning the commandant of the Hamburg shipyard, Admiral Ernst Wolf, wishes them Godspeed and good hunting.

The crew of *U-505* is inexperienced but eager. Only two of the boat's petty officers and one seaman have previous submarine experience. For the next four months, Loewe must whip them into warriors on the water.

Second in command is the chief engineer, Fritz Forster. He oversees the boat's torpedo capabilities and seaworthiness. Forster, who was a year old when his father was killed during World War I, entered the German Navy in 1933 at the lowly rank of stoker, shoveling coal into the boiler, but his technical know-how was quickly recognized, and he was promoted to the engineering corps of cadets. As Loewe gets to know him, he realizes that Forster is a "real pro."

The first watch officer is Oberleutnant Herbert Nollau, who is lucky to be alive. He was serving as radio officer on the heavy cruiser *Blücher* when it was sunk by coastal batteries during the invasion of Norway, suffering 830 casualties.

Second Watch Officer Gottfried Stolzenburg is twenty-nine and a master at navigation. He joined the Nazi Party as member No. 1,540,181 in May 1933.

They soon start clicking as a team. Orders are communicated with simple hand gestures, a flick of the head, or a quick one-word command from Loewe. Spirits are high. Morale is excellent.

The war is going well for Germany. September 1, 1939, saw the invasion of Poland. The next year, Denmark surrendered in a day. Then Norway fell. In May 1940, Hitler turned his attention to Western Europe and, in a single day, established Nazi hegemony over little Luxembourg. Then the Netherlands.

The blitzkrieg cut through Belgium. French capitulation followed. The German Army marched into Paris. Britain was under siege.

And in southern Poland, thirty-one miles west of Krakow, there is a small town, population sixteen thousand. Very few citizens outside its borders are even aware it exists. The Poles call it Oświęcim, but the Germans annexed the town and gave it a new German name, Auschwitz.

There is every reason to believe the Germans will win the war, and the crew of *U-505* are eager to do their part.

The boat is slowly backed out of the pier in Hamburg for her deep-sea acceptance tests. With Loewe viewing the seascape from the conning tower, *U-505* swings her nose toward the mouth of the Elbe and heads for the Kaiser Wilhelm Canal (known today as the Kiel Canal). Entering the Baltic Sea, he links up with twelve other newly commissioned U-boats and engages in joint exercises and training in the tactics of the notorious German wolf pack. Gunnery, torpedo, and test dives follow.

Underwater, *U-505* operates strictly on electrical battery power. This is her great advantage in battle—her capacity to go unseen by the enemy.

Invisibility, however, has its limits. She can move at a top speed of eight miles an hour, for a maximum of twenty-four to fifty hours, depending on conditions, before the batteries must be recharged by spinning the boat's electric motor coils. And that can only be achieved on the surface, when the U-boat runs on diesel fuel.

The winter is bleak. An icebreaker is required to cut a path for *U-505* into the stormy gray waters of the North Sea. The crew is anxious, impatient, and cold.

Suddenly, a crackle. The intercom is coming to life. It's Loewe.

Hier spricht kommandant. "This is the commander speaking. Comrades, the Fatherland is now behind us and our first patrol is underway. *U-505* has been assigned to the Second U-boat Flotilla at Lorient. Our course to France will be around the British Isles. Although we don't have orders to go on the hunt, that doesn't mean we aren't able to sink anything. We'll be hitting heavy weather—a good test for us. But don't let up, because our course takes us through an area under heavy enemy watch. We've had a good start so far. Let's keep it that way." *Das ist alles.* "That is all."

At last, the men of *U-505* know where they are headed: Lorient, a seaport on the west coast of occupied France.

Loewe runs the surface the entire way to France. There is one harrowing moment when lookouts on the bridge raise the alarm. A British destroyer is in the distance, past the windswept troughs of sea. The two enemy vessels take stock of each other, but the British ship fails to launch an attack and *U-505* proceeds onward. The rough seas make it impossible for either side to maneuver into firing position. Game called on account of weather.

Sixteen days after leaving Hamburg, *U-505* arrives in Lorient. German minesweepers escort her into the harbor.

Lorient is peculiar. The red-light district endures, but German engineers and construction crews have converted the rest of the bawdy port on the Bay of Biscay into a massive forward submarine base for the German U-boat fleet. Sixteen bunkers house the submarines. The bunkers—reinforced with twenty-two-foot-thick concrete walls—are impenetrable by Allied bombing.

On a cold and snowy afternoon, a military brass band greets *U-505* as senior staff officers from the Second U-boat Flotilla stand on the snowy pier, decked out in navy blue greatcoats. Loewe recognizes the flotilla chief, Korvettenkapitän Viktor Schütze, a U-boat ace and holder of the Knight's Cross with Oak Leaves. Schütze is a living legend, having sunk thirty-five Allied ships. Loewe pulls into the dock.

"*U-505* reporting as ordered to the Second U-boat Flotilla!" he shouts.

"Heil, crew!" Schütze responds.

"Heil, Herr Kapitän!"

The dignitaries on the pier chant, "We greet the *U-505* and its crew! Hurray! Hurray! Hurray!"

Everyone gathers ashore for a welcoming party. High-ranking officers mingle with the crew. There are toasts to all.

The next few days are a buzz of preparation as *U-505* is outfitted for its first combat patrol.

As a captain who is considerate of his crew, Loewe pays special attention to the provisions. Napoleon once said that an army marches on its belly. The same holds true for the crew of a submarine.

Storing food for fifty men to last one hundred days at sea is, for Loewe, a work of wizardry. An enormous quantity of goods must be loaded on board into impossibly limited space. No nook or cranny can be wasted, and their exact stowage location must be accounted for to maintain the balance of the boat.

The most perishable provision is 3,400 pounds of fresh vegetables, which will start rotting in five days. A supply of two thousand eggs must be consumed within two weeks before they spoil and taste like mothballs, although, according to an old wives' tale, the unrefrigerated eggs will last longer if the crates are turned around every day, but nobody really believes that. For those first two weeks, the crew will feast on eggs until they are sick of them.

Tins of frankfurters are stored behind the diesel engines, not an ideal placement due to the heat from the motors. There are two small bathrooms with toilets to accommodate all fifty men, but necessity requires the bathrooms be converted into storage units. So, the men use the "shit bucket," located in the engine room, when they take a dump—even the captain, although he and the chief petty officer are entitled to use the toilets, which are flushed by a hand pump or air pressure. Unfortunately, the mechanism can sometimes spray right into the skipper's face if the pressure isn't just right. It's safer to avail oneself of the shit bucket, like everyone else. Or they relieve themselves over the side of the surfaced boat, assuming the weather cooperates. They also pee straight into the bilge. The shit bucket is emptied when the U-boat surfaces and some unfortunate sailors are ordered to haul it topside and pour the contents right into the ocean.

There is one refrigerator in the galley, which is large enough for only two other things: a small range and the cook.

In all, *U-505* will set sail with twelve tons of food, including:

- 3,800 pounds of potatoes
- 4,800 pounds of canned meat
- 334 pounds of preserved fish
- 3,400 pounds of vegetables
- 1,200 pounds of dough
- 463 pounds of rice and noodles
- 917 pounds of fresh lemons (to prevent scurvy)
- 550 pounds of butter and margarine
- 611 pounds of soup ingredients
- 400 pounds of marmalade and honey
- 309 pounds of fresh and preserved cheese
- 1,700 pounds of powdered milk
- 154 pounds of strong coffee
- 441 pounds of sugar
- 32 pounds of salt

A large box of prunes, important for good digestion, is available to everyone. And for special treats, 108 pounds of chocolate.

There is also food for the mind. One hundred books make up the boat's library, and eighty phonograph records of popular music are available. To keep himself entertained during downtime, Loewe keeps a copy of Goethe's *Faust* in the captain's cabin.

As *U-505* is loaded with cargo, Hans Goebeler reports for duty. An eighteen-year-old machinist second class with expertise in diesel engines and electric motors, Goebeler will one day hold a special place in the history of *U-505*. But right now, he is low man on the totem pole. Goebeler comes from the small Hessian farming village of Bottendorf. He is an "enthusiastic" member of the Hitler Youth movement; his greatest concern is not death, but that the war will be over before he's old enough to join the fight.

Goebeler stands five-foot-two, an excellent height for submarine service. His small stature will be an advantage in the tight quarters of the U-boat. He has just graduated from the elite submarine school at Wilhelmshaven on the North Sea, and his exhausting journey to Lorient has taken him eighteen hours by railroad. But when he reports to the captain's cabin the next morning, he is fresh, eager—and a bit surprised by the sight of his new commander.

This is a Lion? Axel-Olaf Loewe certainly has a casual way of carrying himself, and Goebeler is thrown off by his informality.

Loewe asks about Goebeler's family and the training he's experienced thus far. Wisely, Goebeler does not utter a single word of complaint about the Third Reich sadists who tortured him at submarine school. He says nothing about the calisthenics he endured at dawn, cleaning the latrines, deep-knee bends until he dropped, or the long, forced marches in snow wearing only shorts. Or getting clobbered in the boxing ring after he was intentionally matched with opponents a head taller.

In many ways, Goebeler is like the other enlisted men on *U-505*: a working-class stiff with a background in the crafts and trades. He will fit in well with the metalworker from the Ruhr Valley, the farmer, coal miner, toolmaker, lathe operator, woodworker, and painter.

Loewe tells Goebeler he will be assigned to the control room. His primary responsibility will be operating the periscope during engagements with the enemy and maintaining the periscope's hydraulic pump. He will also pass commands from the conning tower to the control room when *U-505* is running on the surface.

Goebeler can barely believe it. It's a big job, more than he could have expected. A single slipup, communicating the wrong order, could—

"One small mistake can sink a boat," Loewe says, "but I think you can handle it."

They shake hands. Goebeler snaps to attention and salutes his captain, then finds his bunk and stows his gear.

That first night, Goebeler finds it impossible to sleep. The next night, while everyone else is living it up, he remains in his bunk, rereading training manuals. He is terrified of bungling everything.

The morning of February 11, 1942, the crew assembles on the upper deck of *U-505* to embark on the boat's first wartime mission. It's quite a send-off, with a military brass band and garlands of flowers decorating the bridge.

At 6:00 p.m., Axel-Olaf Loewe shouts the order.

Leinen los!—"Cast off!"

The band strikes up the music. On the dock, a crowd of Navy officers and enlisted men from the Second U-boat Flotilla call out three cheers: "*Hurra! Hurra! Hurra! U-505 ist unterwegs!*" The boat slowly pulls out of the oily waters of the submarine pen and enters Lorient Harbor. Then a minesweeper escorts *U-505* to the outer harbor. As the music from the military band fades, Loewe turns to his executive officer.

"Nollau, throw the flowers overboard."

Sailors are notoriously superstitious. According to German sea lore, bad luck follows any sailing vessel that carries flowers beyond sight of land.

Loewe then orders a deep test dive to 650 feet. All goes well. When *U-505* resurfaces, Loewe signals the minesweeper, "Everything okay."

"Safe return and good hunting," goes the return signal.

The minesweeper heads back to Lorient. *U-505* is on her own.

Once again, there is a crackle. Loewe is about to address the crew over the speakers scattered throughout the U-boat. Up until this point, their mission has been sealed. Now that they are at sea, it is time for the skipper to inform the crew where they are headed.

"This is the *Kapitänleutnant*. We have been ordered to hunt along the West African coast. Our operational area will be the Allies' convoy assembly point off Freetown Harbor. We will be going up against fast ships and it certainly won't be easy. In the meantime, keep your eyes open and be alert! That is all."

The intercom clicks dead.

The crew look at one another. Did they hear him right?

Freetown?

Where the hell is Freetown?

TWO

THIS COLD

STRANGE LAND

New Year's Eve, 1941

T he world is still reeling from the Japanese sneak attack on Pearl Harbor twenty-four days earlier. Commander Daniel V. Gallery would love to be in the thick of the Pacific battleground, but on this day, fate has landed him on the other side of the world, in Reykjavik, Iceland, to take command of the US Navy Fleet Air Base. Hardly a plum assignment.

It doesn't take long for Gallery to assess Iceland. In his opinion, Iceland is a "cold strange land." Put another way, it is a "Godforsaken hole."

Gallery soon realizes that the presence of the Allies is barely tolerated by Icelanders. True, many of them support the Allies, but others are pro-Nazi and see the British and Americans as an occupying force, which, technically, is accurate. The hard truth is that the great majority of Icelanders yearn for neutrality—their historic position in world affairs—and want to sit out the great conflict currently engulfing the planet. Quoting Greta Garbo, Gallery concludes they "vant to be alone."

Gallery was born in Chicago on July 10, 1901. Three of his grandparents emigrated from Ireland during the potato famine. He attended

Catholic schools. Two of his great-aunts are nuns. His brother, John Ireland Gallery, is a priest. The dogma runs deep in his family.

When Dan Gallery entered the US Naval Academy at age sixteen, he stood five-foot-six, weighed a scrawny 127 pounds, and had big ears. But very few plebs messed with him because he was also a resourceful wrestler, bantamweight division. He would compete in the 1920 Olympics in Antwerp. The midshipmen called him Diz, Dizzy Dan, Dizzums, and Irish.

Gallery graduated Annapolis in the top 10 percent of the class of 1921, but the early days of his naval career were far from glorious. On the battleship *Delaware*, Ensign Gallery shoveled coal. Then he was off to the USS *Arizona*, the cruiser *Pittsburgh*, and the battleship USS *Idaho*. In 1926, he transferred to the Navy's flight school in Pensacola and learned to fly. Within a year, six of his twenty aviation classmates had lost their lives in crashes, but it seems that the luck of the Irish held for Gallery—he came through it without a scratch.

Gallery met his wife, Vera Lee Dunn, on a blind date in Washington, DC. Vee was no delicate society debutante or traditional Naval Academy trophy wife. She was, of all things, a US deputy marshal. But Gallery thought her very beautiful. He broke the seemingly good news in a "Dear Papa" letter to his father, Daniel V. Gallery Sr., a Chicago lawyer.

"I think I'm going to get married . . . to Miss Vera Lee Dunn of Washington. Miss Dunn is a fine girl and I'm sure you'll like her when you get to know her." That wasn't the only bombshell Dan Jr. dropped:

"She is left-handed, from the north of Ireland, but she is going to become a Catholic."

The news profoundly upset his father—a tough son of a gun—as it did Gallery's brother, who objected on religious grounds as well as Miss Dunn's "lack of social status." Whoever heard of marrying a deputy marshal? Dan Jr. was willing to butt heads with the old man and Father John, to whom he wrote, "I wouldn't have the crown princess of England if I could have Vee instead." Dan and Vee married in 1929 and ended up adopting three children. They lived on a farm outside Fairfax, Virginia, which they named Harmony Farm.

Dan Gallery is forty when World War II breaks out.

He feels that his time has come at last. He knows ships, and he knows how to fly. He has all the credentials to be given command of an aircraft carrier, which makes his assignment to a land base in the Arctic boondocks so vexing.

Right from the get-go, Gallery finds himself in a dicey political situation. He has to report to five bosses in all—two US admirals, a US general, a British Royal Navy admiral, and the RAF commodore. No man can serve two masters. As Gallery likes to point out, it's right there in the Bible. Serving five is nuts.

The British officers are also skeptical of Gallery. One look at his mug told them he must be Irish. The rugged face with the impish crooked smile looks like a relief map of the Emerald Isle. During an otherwise friendly conversation, Air Commodore Hugh Pughe Lloyd wonders whether Gallery is the right man for the job, given he has to liaison so closely with the British military. After all, England is the historical oppressor of the Irish people.

Nothing of the sort, Gallery assures Lloyd.

"In fact," Gallery tells him, "I'm eternally grateful to your ancestors for persecuting my ancestors so that I was born in the USA."

What balls! Luckily, Lloyd gets the joke.

All the senior RAF and Royal Navy officers that Gallery is dealing with seem to have three names and a bewildering cascade of initials following their signatures: D.S.O, K.C.B., C.B.E., and so on. So very British. Gallery has no idea what any of it means. The only initials after his name are *Jr.* For the hell of it, Gallery starts using *D.D.L.M.* One day, Commodore Lloyd stares at a memo from Gallery concerning an operational matter. He looks perplexed.

"I didn't know you Americans had any such thing," the commodore says.

Gallery can see the wheels churning in Lloyd's head. Lloyd asks, "Just what does *D.D.L.M.* stand for?"

Is another Gallery quip about to be delivered? You betcha.

"It means, 'Dan Dan the Lavatory Man.'"

Lloyd has a big laugh over that one. Fact is, they see eye to eye on just about everything having to do with the war, including the occasional humor to break the tension.

But Gallery's latest assignment is serious business. In just about the worst weather in the world, Gallery has been ordered to hunt and kill Nazi submarines prowling the North Atlantic. Hitler's fleet of U-boats is threatening to starve England into submission. The stakes are as big as they can get. The sea lanes must be kept open and the Allied ships carrying food and war supplies to the British must get through.

In these early days of 1942, the Allies are losing the Battle of the Atlantic.

Iceland offers a miserable existence for American military personnel. They walk ankle-deep in mud and live in primitive barracks as howling Arctic winds burrow into their bodies. And, by the way, Navy chow is nasty, and the US base outside Reykjavik is infested with rats.

Gallery knows that to build an effective fighting force and keep his boys from going stir crazy he needs to provide adequate shelter. First order of business is to deal with the rats. Gallery puts up a bounty: a dollar per head. The rifles come out and everyone starts taking pot shots. Soon the rat population is exterminated. Great, but they didn't come here to kill rats.

Next, Gallery orders Navy Seabees to construct new barracks as well as a recreation hall with a gym, basketball court, and movie theater. Quonset huts—prefabricated structures made from corrugated steel—are erected. There are fourteen bunks in each hut, plus card tables, writing desks, easy chairs made from shipping crates, and a phonograph and radio. The walls are decorated with pinups cut out of the pages of *Esquire* magazine. Permission is granted to date the Army nurses working at US military hospitals in Reykjavik.

Every night is movie night. The base receives only ten first-run Hollywood films a month, flown in from the States, so several screenings are repeats. The men don't mind. They still pack the movie theater every night. The most popular film is *Boom Town*, about the early days of oil wildcatters, starring Clark Gable and Spencer Tracy. If no new release is available, they keep showing *Boom Town*. In no time, all the guys learn the *Boom Town* dialogue by heart and recite the lines during showtime. They really have a hoot when they get to the hot love scenes with Hedy Lamarr.

In return for fostering such goodwill, Gallery has but one request, which he condenses into a single word: "kwitcherbellyakin." (Translation: "Quit your bellyaching.")

But Gallery himself struggles to keep from complaining. By the summer of 1942, after six months in Iceland, he's ready to blow. He feels the top brass in Washington has forgotten they exist. He's had it with softball tournaments and all this monkey business that has nothing to do with victory over the forces of fascism. It's one thing to create an esprit de corps. That's just good leadership. But, really, what about the business of sinking Nazi submarines? His men have yet to destroy a single U-boat, and it's infuriating. Three confirmed sightings in the time Gallery's been here—all squandered due to lousy luck and inattention to minor details.

Gallery calls the pilots together and reads them the riot act. And, by the way, he informs them, the officers' club is closed until further notice. Get your first kill, *then* we'll talk about opening the officers' club again.

The no-more-Mr.-Nice-Guy speech works, as things finally turn around on August 20, 1942.

It is 5:01 a.m.

US Lieutenant Robert "Hoppy" Hopgood and his crew of seven men are flying a lumbering PBY Catalina, an amphibious aircraft, north of Iceland. Their assignment is to provide aerial support for a Royal Navy task force en route to a mine-laying operation in the Denmark Strait. Conditions are poor: The sky is overcast, with heavy winds and rain squalls. About fifty miles from linking up with the task force, Hoppy sees what sure looks like a steel cigar riding the surface. He descends to an altitude of five hundred feet. Confirmed! It's a U-boat! He can even make out the German crew taking in big gulps of fresh air on the deck.

Hoppy launches an attack run. He opens the bomb bay doors and drops his entire payload of six depth charges. The aim is so precise that three of the bombs actually land on the deck, but don't explode because they are set to detonate at twenty-five feet underwater. Then a remarkable thing happens. The Germans dump the depth charges over the side. Sure enough, the bombs sink twenty-five feet down. Boom! *Boom-boom-boom!* The Nazis inadvertently complete the destruction of their own vessel.

From his Catalina, Hoppy can see the blasts have crippled the U-boat. Unable to submerge, she's limping along the surface, leaking thousands of gallons of oil. Hoppy opens fire with machine guns to let the Germans know who's boss.

Back in Reykjavik, Gallery and his officers gather around the radio. For the next three hours, they listen to Hoppy giving a play-by-play account of the engagement.

When British destroyers reach the stricken vessel, she is identified as *U-464*. Of course, by the time the Brits board and take possession of the enemy boat, the Germans have thrown all the code books and other important documents over the side and destroyed the Enigma machine. Contained within a wood box measuring 13.5 inches by 11.6 inches, the Enigma machine has a keyboard consisting of the twenty-six letters of the alphabet. You hit a key and a light goes on. And if you open the back of the machine you see a crisscrossing entanglement of wires that look like spaghetti and four electromechanical rotors that turn with every stroke of a key. The Enigma encryption machine scrambles ordinary words into gobbledygook. There are billions—*sextillions*—of potential combinations, which is why the Germans consider the machine unbreakable, unless, of course, you have the code books.

U-464 may be in Allied hands, but its greatest secrets remain a riddle.

Hoppy ends his radio report with this: "Personal for Commander Gallery—Sank sub, open club!"

Finally! That night, the American lads party well into the midnight sun. Gallery puts Lieutenant Hoppy in for the Navy Cross. It's not every day a lone airplane leads to the capture of a German U-boat.

Once again, the officers' club becomes the hub of social activity for Gallery and his senior staff. They chew the fat around the fireplace and read the big city papers flown in from the States, only a day or two old, so they're totally up to speed on the state of the war and news from the home front. There's an interesting headline out of Washington: Henceforth, every bottle of milk produced in the USA will be inscribed with the slogan, "Buy War Bonds." They also read that the rationing of meat is looming and restaurants are being urged to adopt "meatless Tuesdays." At Fenway Park in Boston, 38,814 fans watch the world champion New York Yankees split a doubleheader with the Red Sox. And in Berlin, Nazi Minister of Propaganda Joseph Goebbels warns the Allies that might, not piety, rules the world. What a guy.

One story around the fireplace Gallery finds particularly fascinating: the capture of *U-570*. It happened before his arrival in Iceland, and it is a very odd tale.

U-570 was on her first war patrol, and it was a cursed voyage. Everything started going wrong when a seasick rookie sailor started puking his guts out into a bucket. Nothing induces vomiting more than seeing another guy throwing up. Pretty soon it seemed as if a third of the crew of *U-570* was disabled by seasickness. Then the air compressor broke down.

Three hundred miles south of Iceland, in choppy waters, the skipper of *U-570*, Captain Hans-Joachim Rahmlow, gave the command to surface and air out his stinking boat. Wouldn't you know it, a British Hudson bomber was flying directly overhead and saw *U-570* pop out of the water. The RAF pilot flew in low, made a beeline for the submarine, and dropped his payload of depth charges. One detonated just ten yards from the submarine. The explosion slammed *U-570*, and she almost rolled over before settling. Power was knocked out and the boat started leaking. In the engine room, the batteries mixed with salt water and released a gas, which the terrified crew thought was chlorine. The captain issued the order to abandon ship. On deck, a gale was blowing, making it impossible for Rahmlow to scuttle the sub and escape on lifeboats.

Flying above, the British pilot couldn't believe what he was seeing— the Germans were waving white shirts in surrender. Also, their underwear! So much for the "master race."

It took a full day for a Royal Navy tug and destroyers to catch up with *U-570*. Rahmlow meekly accepted a tow to Iceland. Of course, by the time the Brits boarded her, all the secret codes had been ditched over-board and the Enigma machine smashed to pieces, just like *U-464*.

Listening to this account at the fireplace, Gallery gets to thinking. The story tells him a very important lesson, which is that when German sailors are faced with dire circumstances, they do not show a willingness to die for their Führer, the Fatherland, or godless totalitarianism. Just like any young man in war, they have a strong instinct for self-preservation.

U-570 . . .

U-464 . . .

It's all coming together in Gallery's head.

What-if scenarios start bouncing around the fireplace.

One of Gallery's officers wonders, "Why can't we board and capture a sub with one of our PBYs?"

More what-ifs follow.

What if you could figure out a way to cripple a U-boat so she couldn't submerge? What if you were able to board a submarine *before* the crew scuttled the vessel and destroyed the code books and the Enigma machine?

The lively discussion goes back and forth. A bunch of motivated young American warriors spinning stories, killing time, chewing the fat.

Gallery calls it a night and steps outside the officers' club. He looks up. In the heavens above, he observes a breathtaking aurora borealis.

Capturing a German submarine . . . with all its secrets and technology intact.

Was it even possible? Could it actually happen?

Staring at that brilliant display of swirling polar lights, somehow the idea doesn't seem so crazy.

THREE

KILL! KILL! KILL!

Kapitänleutnant Axel-Olaf Loewe swings *U-505* out of the Bay of Biscay off the coast of France at flank speed. Night is falling, and Loewe uses the cover of darkness to cross these treacherous waters. British ships and planes are on patrol, hunting for Nazi U-boats departing Lorient, and Loewe's lookouts are on full alert. The sooner he clears the Bay of Biscay and sails into open Atlantic waters, the easier he'll breathe.

The next morning, breakfast is announced over *U-505*'s loudspeaker: *Aufbacken!* "Chow's up!" Strong coffee is served, with buttermilk soup, biscuits, hard bread with butter and honey, and, of course, eggs—as many as they can stomach. Already the crew is growing sick of them. The cook fears he'll soon be dodging eggs.

The hours pass. *Mittagessen ist fertig*—"Lunch is ready." Soup, potatoes, canned meat, vegetables, fruit.

More hours pass. *Abendessen ist fertig*—"Dinner is ready." Sausage, cheese, bread, coffee, chocolate.

It's a routine repeated daily for three straight weeks as *U-505* sails south-southwest, and the monotony is getting to everyone.

When he's off duty, Hans Goebeler keeps occupied playing chess or cards. Trivia games are a favorite diversion. History is Goebeler's best subject, and he's a whiz at naming the capitals of foreign nations. He often wins and the award is an extra scoop of dessert—a concoction of crushed ice and raspberry pancake syrup.

Passing the Azores, Loewe orders one of the diesel engines shut down to conserve fuel. *U-505* is running at half-speed when it crosses the Tropic of Cancer and the weather turns blissfully warm. But the rules for enjoying it are strictly enforced. Due to the ever-present potential of a crash dive, only two or three sailors at a time are permitted on deck to relax and enjoy a smoke and fresh air. (Smoking is forbidden inside the boat due to the risk of detonating the storage batteries.) The deck is made of soft pine wood treated with black tar as a preservative as well as camouflage against enemy aircraft. There's a backup of men waiting to go topside, standing in line by the ladder awaiting their turn. No one is permitted to stay longer than the time it takes to finish one cigarette. It's a short but blissful break.

They've changed into tropical gear: canvas shorts or gym trunks, no shirts, or a net shirt, and pith helmets with large brims. Goebeler enjoys the breeze warming his face. Back home in Germany, this time of year, he knows his family is probably huddled around the fireplace.

In no time, Goebeler and the rest of the crew have chocolate-brown tans.

What a relief it is to feel that sunshine! If only they could stay topside all day long.

Because inside the boat, it's like a furnace. The temperature in the engine room exceeds a hundred degrees. A steady drizzle of condensation water drips onto their faces. And the stench! For Goebeler, it smells like a "fetid stink of rot and decay—a cold clammy coffin." There are only thirty-five bunks, so they must be shared. As soon as your shift starts, you jump out of your bunk and a crewmate hops in to take your place. It's called "hot rack," or hot bunk because the bunk is in use twenty-four hours and the mattress reeks from body odors. Only the officers and chiefs have their own cots.

On March 8, 1942, *U-505* at last arrives at its designated patrol zone, 210 miles southwest of Freetown, Sierra Leone.

She is riding the surface. Four lookouts are standing on the bridge continuously scanning the horizon with their high-grade Carl Zeiss binoculars. They are each assigned a sector. On the starboard side, one sailor sweeps from 0 to 90 degrees. The second man scans from 90 to 180 degrees. The same duty is repeated by two other lookouts on the port side, for total 360-degree surveillance.

The lookouts have been selected for their perfect eyesight and night vision—a natural ability to see in low-light conditions. Talking is strictly forbidden. And no smoking. Daydreamers are disqualified. A single moment of distraction—blowing your nose, scratching your crotch, cleaning the lens of the binoculars—can lead to a missed opportunity. Or worse, failure to raise the alarm in time if an enemy plane is about to attack. To stay sharp, the lookouts are replaced every four hours with fresh personnel.

It is 6:36 p.m.

The day has been spent in yet another vain search for Allied vessels. Suddenly, a lookout spots something. *There!* On the horizon. A plume of smoke. It's a steam freighter, zigzagging, with lights dimmed, running unescorted at twelve knots, on a course to Freetown. Must be British. Maybe American.

Straightaway, all those weeks of tedium turn into a frenzy of activity that consumes every member of the crew. This is what they have been training for. Loewe orders a course correction and flank speed at eighteen knots. It takes *U-505* four hours to catch up to the freighter and maneuver into attack position.

Tauchen! Loewe calls out. "Alarm! Dive! Dive!" Thirty seconds later, *U-505* is twenty feet below the surface, standard periscope depth. She is now invisible. The crew is at battle stations. The tension is unbearable.

"Make ready one and two," Loewe orders.

There are four forward torpedo tubes, two on the port side, two on the starboard side.

The data is fed into the torpedo indicating course, range, speed, and gyro angle.

When the torpedo doors are open, the torpedo tanks flood with water.

In the control room, Hans Goebeler, his heart hammering with excitement, adjusts the boat's trim to compensate for the flooded torpedo tubes. It is his responsibility to keep the submarine level.

Silence.

The freighter is six hundred meters away.

U-505 is ready to attack.

Torpedoes los! Loewe calls out. "Torpedoes away! Fire one! Fire two!"

Whoosh! Another *whoosh!*

It's a two-torpedo spread, setting off for the freighter.

The seconds pass.

Then, nothing.

What is happening?

They can't believe it. Both torpedoes—a miss and another miss.

Loewe betrays nothing, not irritation or disappointment. Zero emotion. There is still time for another go at it if he can keep his head. Focus! His face remains stoic, but he's thrilled on the inside with a new plan. *Yes! That's what we'll do!* He orders a change of course at top speed. They're going for another shot from a different angle. Hans Goebeler's eyes glint. He has just one thought in this mind: *That damn freighter.* He doesn't know her name, nation of origin, what cargo she carries, or the passengers and crew on board. All he knows is that this is war, and *U-505* must sink that ship.

New orders are communicated to the weapons crew. Once again the data is fed into the torpedo's mini-computer: Distance: four hundred meters. Depth: three meters.

Loewe barks: *"Torpedo—los!"*

Whoosh!

Calculating the distance at four hundred meters, the Germans figure it'll take twenty seconds for the eel to reach its target.

They start the count: "fifteen . . . sixteen . . . seventeen . . . eighteen . . . nineteen . . ."

Boom! Even underwater, they can hear the muffled explosion.

Loewe barks out another order.

Auftauchen. "Surface."

The klaxon sounds. *AH-OOH-GA.*

They now get a good look at the stricken enemy freighter. It's a direct hit. She is stopped dead in the ocean with a hole amidship. Then the radioman on board *U-505* picks up an SOS signal.

She's British, all right. Her name is the SS *Benmohr.* She is out of Durban, South Africa, on her way to Freetown for refueling. Her ultimate destination is Oban, Scotland. She is carrying a cargo of war supplies: rubber, silver bullion, and pig iron for the manufacture of steel.

The freighter is listing. Loewe can see the British crew abandoning ship and going over the side in lifeboats. Her stern is high out of the water. She is definitely a goner.

U-505 creeps closer. At three hundred meters, Loewe orders his submarine to come to a halt. All fifty-six men of the *Benmohr*, including the ship's master, David Boag Anderson, are now safely on board the *Benmohr* lifeboats.

It's time for the finishing blow. Loewe cannot risk the British Navy salvaging the *Benmohr* or recovering her cargo. He orders one more torpedo fired—the fourth of the skirmish. From this distance, he can't miss. The torpedo hits the keel and a long death slide follows as *Benmohr* sinks to the bottom of the sea. Then *U-505* makes a hasty departure before British forces can respond to the SOS distress call.

U-505 is on the scoreboard with her first kill: a fully loaded British freighter.

The Germans all agree they are off to a good start, but they also know that they've put themselves in even deeper danger. The Allies will seek vengeance.

Sure enough, the very next day, a plane is spotted overhead. It's a British Sunderland, also known as a flying boat, although the Germans call them "bumblebees." No matter what it's called, the main thing is that it's a bomber on the hunt for the U-boat that sank the *Benmohr*. Loewe orders *U-505* to dive, and only when enough time has passed does he deem it safe for his submarine to surface again. Then, they spot something. There, on the horizon, in a howling rain squall, a faint plume of smoke. Looking into his binoculars, Loewe ascertains that it's a large tanker. Like the *Benmohr*, she's running zigzags, only with more yawning this time—further evidence that all Allied ships operating in the vicinity have been alerted that there's a U-boat on the prowl. Because the tanker is not flying a flag, Loewe reckons they are trying to conceal its nation of origin. A quick skim of the *Merchant Ship Recognition Manual* shows that she is British designed, about eight thousand tons, considerably larger than the *Benmohr*. She is lying low in the water, apparently fully loaded

KILL! KILL! KILL! | 23

with oil, probably destined for the British Army fighting Erwin Rommel, the Desert Fox himself, in North Africa, and so a prized target for *U-505*.

Loewe ponders the possibilities. He is about forty-five miles from where he sank the *Benmohr*. Tactically, he's facing an interesting situation. He can chase the vessel—but why waste fuel? Better yet, he can let the tanker come to him. As of now the tanker is pretty much on a direct path to *U-505*. Loewe doesn't have to do much except stay put under the sea and wait.

Loewe orders *U-505* to dive to periscope depth. She is hovering about thirteen feet below the surface. Just waiting, hanging tight.

Two hours pass. It is 11:31 in the morning when Loewe peers in the crosshairs of his periscope and sees the tanker is dead ahead, just two hundred meters away, in the direct path of his front torpedo tubes. *Whoa!* Anything closer and Loewe risks damaging his own submarine when he opens fire.

In the control room, Hans Goebeler's stomach is churning. What the hell is the skipper waiting for?

"Torpedo 5—Los!"

"Torpedo 6—Los!"

It's a double shot.

The crew counts the seconds.

"... seven ... eight ... nine."

They're so close to the tanker they don't get past the number nine.

Boom!

A moment passes, and then, suddenly, a monster shock wave hits *U-505*, knocking every German off his feet. It can only mean Loewe's analysis is correct—the tanker is fully loaded with gasoline. But he never anticipated this ricochet.

For the moment, Loewe can't see anything through the periscope. Huge waves are obscuring his vision, and he must wait two minutes for the ocean to settle down. The blast has even cracked *U-505*'s diesel engine clutch, but not beyond repair. It could have been worse.

When Loewe can look through the periscope again, he is taken aback because there's no sign of the tanker. It no longer exists. All that's left is a giant red ball of flames and thick clouds of black smoke.

He barks, "Battle stations!"

The crew scrambles.

Loewe issues another command: "Surface!"

The hatch pops open. Loewe and his officers climb the ladder to the bridge and behold a stunning sight. Before them lies an immense spreading oil slick in the shape of a ring, littered with floating debris and wreckage. Twenty-four dazed survivors are bobbing around in this mess, their faces blackened with oil. They can't believe the fate that has befallen them.

The tanker is called the *Sydhav*. It turns out she's not British after all—she's Norwegian. She left the Dutch island of Curacoa in the Caribbean on February 11, bound for Freetown with a crew of thirty-six men and 11,400 tons of oil for the Allied war effort. The first torpedo hit *Sydhav*'s engine room, the second torpedo a little forward of it. There was no time to launch any lifeboat. The tanker was pretty much obliterated. The men jumped overboard but twelve of them were pulled under by the suction of their sinking vessel, including the ship's master, Nils Helgesen, of the Norwegian merchant fleet. He went down with his ship.

Those oil-soaked survivors include mechanics, carpenters, firemen, oilers, stewards, cooks, and radio operators, and their problems are far from over, because they are floating in shark-infested waters. Their only hope is to reach two rafts and an overturned lifeboat. One merchant sailor sprawls himself across a mattress, thinking himself safe, but his legs are dangling over the side—and the jaws of a man-eating shark drag him away. A thirty-one-year-old Chinese cook from Shanghai named Woo King Sung is afloat, clinging to debris. From the bridge of *U-505*, Kapitänleutnant Loewe questions him but doesn't get much information. By nightfall, Woo King Sung will be dead from burn injuries.

Loewe is prepared to give the *Sydhav* survivors food, water, morphine, salve, and bandages but a lookout on *U-505* raises the alarm. There, in the eastern sky, he eyeballs a British Sunderland flying boat coming right at them. The bomber is eight thousand meters away. Nothing Loewe can do now except order an emergency dive. The survivors of *Sydhav* must fend for themselves.

The irony of it! Their chances for survival would have been better if a plane from their side had not shown up.

Now safely submerged, Loewe settles back in his cabin. What momentous times he is living through. He marvels at the day's events, but he must ask himself a troubling question: Was the *Sydhav* a legitimate target? From Loewe's point of view, he was justified in attacking her without warning because the tanker was engaged in commerce in support of the enemy. Under international laws of war, he also had an obligation to search for survivors and render humanitarian assistance. He certainly intended to help—that is, until the British bomber came at him and he had to make that sudden dive. Even so, in his daily diary entry he notes that the tanker "sank in 25 seconds." It'll look better on the official record. As to the name of the tanker, Loewe lists it as "unknown." *Sydhav* never appears in *U-505*'s war journal.

The Brits are ticked off. Two Allied tankers on their way into Freetown—destroyed in two days! This crisis must be dealt with, and the decision is made that, henceforth, no Allied merchant ship is permitted to proceed out of Freetown without a Royal Navy escort. Air patrols also intensify.

The crew of *U-505* know they, the hunters, are now the hunted ones. The skies above are dotted with British "bumblebees"—bombers scanning the ocean for the submarine. Loewe is compelled to remain out of sight underwater for as long as his batteries last, frittering the hours away holed up in the stinking steel cigar.

Good fortune at last comes his way in the form of a violent tropical rainstorm. For the time being, the British air fleet is grounded. Loewe takes the opportunity to surface and recharge the batteries during the rain squall. The crew also reloads the forward torpedo tubes.

Loewe radios Admiral Karl Doenitz, the "Great Lion" of the Kriegsmarine. Informally, Doenitz is known to his men as "Uncle Karl," but the kindhearted nickname disguises a fervid Nazi who is the architect of Hitler's wolf pack submarine strategy. His grand exhortation to his fleet of U-boats is straightforward: "Kill! Kill! Kill!"

Loewe requests permission to depart the tropical waters of West Africa and sail for South America, where enemy ships are abundant, and he would be out of the reach of the British bumblebees. Doenitz radios back. Negative. Remain in your current patrol area.

On April Fools' Day, Loewe decides to change course for the equator, for no military reason whatsoever other than to offer an entertaining diversion for his frustrated crew.

Every sailor in the world knows that when you cross the equator for the first time, tradition calls for a hazing ritual. It's an old maritime rite of passage meant to pay tribute to Neptune, the king of the sea.

All the men of *U-505* gather on the deck. A sailor dresses as Neptune in a flowing white beard and robe and presides over the ceremony. To his utter mortification, a young sailor with a baby face is chosen to portray Neptune's mermaid wife. But that's nothing compared to the hazing ritual. It is, to put it mildly, a doozy. Laxatives made from flour and castor oil and the size of golf balls have been prepared. No chewing is permitted. The rookies must swallow the balls whole. Then they are told to pull down their shorts and crawl out to the end of a long wood plank extending from the side of the submarine and sit on their heinies inside a large cutout hole where they wait for nature to take its course. The Germans sure love their scatological humor.

The next day, the romp at the equator is over and it's back to chasing down Allied ships. No longer does it seem as if *U-505* has the ocean all to herself. A new target comes into view. She's a tanker and, in keeping with the new Allied defense strategy, a British ship is sailing close by as escort. Loewe decides to wait for the camouflage of night before launching the attack.

April 3, 1942.

21:50 hours.

Loewe opens fire with two torpedoes from a range of one thousand meters. Both miss. No surprise considering the distance. The Allied crew is oblivious to how close they came to being blown out of the water. Loewe dares not draw any nearer with that escort breathing down his neck. For the next ten hours, Loewe stalks until he loses his target in heavy squalls, then finds her again. He prepares to submerge and launch an attack—but wait! Coming in the other direction, his lookouts spot a more intriguing target—a steamer, escorted by *two* British naval ships. This can mean just one thing—she's carrying valuable cargo. Loewe doesn't have to give chase—the steamer is not only coming right at him, she's on a collision course.

It's a gift! *U-505* dives to periscope depth and waits. The crew braces. Hans Goebeler, in the control room, marvels at Loewe's "nerves of steel." Observing his target through his periscope, Loewe sees that the steamer isn't zigzagging, which is a standard defense maneuver in these waters. She's probably confident the two escorts will keep her safe. Truth is, she's a sitting duck. Loewe lines up *U-505* for an easy kill. At four hundred meters, he fires two torpedoes. The first eel misses the mark—too wide or too deep, he can't tell. But the second eel nails its target on the port side, exploding with a terrible roar, slightly forward of the bridge. A hole twenty-by-eighteen feet blasts open just forward of the bridge. Ten African stevedores who were relaxing on the deck are killed instantly.

An SOS message is broadcast, and as Loewe's radio operator monitors the distress signal, he learns the identity of the ship *U-505* has just blown up. She is called the *West Irmo*. She's an old-fashioned steamship, built in 1919, but her long years at sea are now coming to an end.

The master, Norwegian-born Torlief Christian Selness, issues the order to abandon ship. The *West Irmo* is sinking fast, and in minutes, she is gone. Loewe can hear the boilers exploding on the way down.

There are ninety-nine surviving merchant seamen and longshoremen.

Two miles away is *West Irmo*'s escort, the HMS *Copinsay*. The skipper, Edward Robert Harris, sees the horror playing out. He steams for the lifeboats and hauls all ninety-nine *West Irmo* survivors aboard. Later, not now, he'll have some hard questions for Selness, first among them—why wasn't he zigzagging?

Right now, he'd love to have a crack at the U-boat that did this damage. The *Copinsay* is armed with thirty depth charges—"wabos," as the Germans call them—and they can kill or at least seriously damage *U-505*. *Where is that damn U-boat?*

Loewe figures it's time to turn tail. No point in sticking around and picking a fight. His mission is to destroy merchant ships, not warships that can fight back. *U-505* has done its job and lives to fight another day.

So far, 1942 is emerging as a terrible year for the forces of democracy in their struggle against Nazism and the U-boat menace. The numbers tell the grim story: By the end of the year, the Allies will have lost 1,570 ships and the Germans only 86 U-boats. Off the East Coast of the United States, an extraordinary 105 U-boats are in operation. A total of 524 Allied ships are sunk between July and December within the waters of the American security zone.

America and Great Britain are on the brink.

FOUR

THE EYES OF TEXAS
ARE UPON YOU

Wayne "Mack" Pickels Jr. is preparing to go to war. His choice of service surprises everyone who knows him.

Not long ago, Pickels was the star quarterback and captain of the Alamo Heights High School football team in San Antonio, Texas—number 17. He also played varsity baseball and basketball, and he was sports editor of the yearbook. He's five-foot-ten, solidly built, weighing 205 pounds—none of it fat. He's a good-looking young man with a fair complexion, blond hair, and brown eyes. He lives with his parents and fourteen-year-old kid brother, Toby, at 647 Wheaton Road in the town of Fort Sam Houston.

Yes, he's an Army brat, all the way. His father, Wayne Sr., is a career officer in the Army's quartermaster corps who served under General Pershing in pursuit of the Mexican revolutionary Pancho Villa in 1916. Pickels Sr. also fought in the Battle of Verdun during World War I.

Wayne Jr.'s upbringing was as wholesome and all-American as anything Norman Rockwell could ever paint for the *Saturday Evening Post*. He used to go door-to-door selling *Liberty* magazine and earned enough coupons to buy a pocketknife, compass, and baseball glove. In high school, he belonged to Hi-Y, the social club for high school boys affiliated with the YMCA. One summer he dug trenches for sewer lines. The job served two purposes: It earned him thirty cents an hour, and it conditioned him for the upcoming football season.

Pickels was a true gamer on the field, refusing to quit, no matter what. In his last season with the Alamo Heights Mules, he had to play with a heavy brace on his right knee, the result of an injury sustained in the final five minutes of the previous season's closer against archrival Lanier High. He graduated high school in May 1941 and enrolled at Southwest Texas State Teachers College.

Three months later, the Japanese bombed Pearl Harbor, and the United States entered World War II. Of course, this lifelong Army brat rushes to enlist—in the Navy.

Nobody saw *that* coming! Wayne Jr. was always expected to follow family tradition and join the Army. The fact that he signs up for Navy duty is such a shock it even makes the hometown newspaper, which had been reporting on young Pickels's football feats for two years.

"You wouldn't expect an Army officer's son to join the Navy but Wayne Pickels Jr. did just that," notes the *San Antonio Light*, under the irreverent headline, "Howz That?"

When he hears the news, Pickels's father is a little ticked off, but gets over it quickly. He respects his son's decision, especially considering what happens on March 1, 1942.

That's the day the heavy cruiser USS *Sam Houston* is sunk in a torpedo attack by the Japanese Imperial Navy during the Battle of Sunda Strait. Of the crew of 1,061, only 368 survive.

When the fate of the *Houston* is made known, a call goes out for volunteers to replace every sailor killed in the sinking. One thousand Texans step forward to avenge the loss of the ship and its fallen sailors. The new recruits are called the Houston Volunteers. As their ancestors had once rallied to the cry of "Remember the Alamo," now these young Texans are heeding a new battle cry: "Remember the *Houston*."

On Memorial Day 1942, as a crowd estimated at two hundred thousand watches, the volunteers line up on Main Street in downtown Houston and are administered the Navy Oath. President Roosevelt sends a message of gratitude: "Every true Texan and every true American will back up these new fighting men."

Mack Pickels is one of the volunteers. He can be seen in a photo that runs on front pages across the United States, standing in the front row, a small, bulging suitcase in hand, ready to head off to war. He and the other inductees march to Union Station where they board trains to begin their military service.

Four days later, the trains pull into San Diego and the men report for duty at the United States Naval Training Center. A band is there to greet them, playing "The Eyes of Texas Are Upon You." They are ready to serve.

But what a motley collection of Texans it is. Flabby old guys, shoulder to shoulder with sinewy teenagers who look as if they've ditched the senior prom to be here.

"We are proud to welcome you into the Navy," says the commanding officer of the nation's largest training base. "Be ready to do the job you pledged your fellow men in Houston to do."

They respond with thunderous cheers.

Apprentice Seaman Pickels spends the next ten weeks in boot camp. There are lots of marches and small arms drills, loading large-caliber guns and learning the basics—Navy customs, shipboard structure, knot tying, line splicing, and the use of semaphore flags. He and the other lads perform endless calisthenics and run the obstacle course until they're ready to drop. No liberty is permitted. When off duty, Pickels hand-washes his uniforms, cleans the barracks, and stands guard duty. The pay? A buck a day.

His mother writes frequently, and, following Navy advice to all parents, keeps the letters cheerful. His father—recently promoted to Army colonel and serving in North Africa—always ends his letters, "Have faith in the Lord."

Regular promotions for Pickels quickly follow: Seaman Second Class, then Seaman First Class. Pickels completes gunnery fire control school.

Then he receives his first assignment—Pearl Harbor and the destroyer USS *Perkins*.

He's thrilled, until he learns his duty on board the *Perkins*—cleaning meal trays in the mess hall.

One day, general quarters is sounded and Seaman Pickels races to his battle station in the Fire Control Section. *This is what I signed up for!*

Scooting down a ladder, he makes a clumsy turn. His right knee buckles under him, aggravating that old high school football injury. He's sent to sick bay and then transferred to the Navy hospital at Pearl Harbor where the surgeons do a good job fixing his knee.

A wounded sailor before he's even seen action. Bad luck. Or is it good fortune in disguise?

As Pickels recuperates, the *Perkins* is dispatched to the South Pacific where she is engaged in the shelling of Japanese forces holed up on Papua New Guinea. Then, a dreadful setback—the *Perkins* collides with an Australian troopship, splits in half, and sinks. Nine sailors lose their lives. When Mack Pickels hears the news, he feels terrible for his shipmates. What forces, he wonders, are at work? Was it divine intervention that put him in the hospital instead of on the ship? Was his life spared for a reason, perhaps for some higher calling?

Christmas 1942 finds Pickels still in Hawaii, with which he is quite familiar, having spent several years of his childhood at Schofield Barracks when his father ran the bakers and cooks school.

An elaborate Christmas dinner is prepared for the sailors, starting with shrimp cocktail, followed by cream of tomato soup, Hawaiian baked ham, and candied yams. For desert: ice cream and mixed nuts.

Soon after the new year, Pickels is transferred to another ocean and finds himself in the thick of the Battle of the Atlantic, promoted to coxswain, and assigned to the destroyer escort USS *Pillsbury*, escorting convoys from US ports to Gibraltar and Casablanca.

Aboard the *Pillsbury*, he meets Zenon Lukosius, the twenty-four-year-old son of Lithuanian immigrants. His buddies call him Zeke, or Luke. His father was killed in a streetcar accident when he was fourteen, and Lukosius quit school to help support his family. Before the war he made eight dollars a week working for the Cincinnati & Eastern Railroad and gave seven dollars of his salary to his mother, keeping just one for himself. So, why the Navy? For one thing, his big brother served in the Navy. For another, he wanted to see the world. To paraphrase Irving Berlin, Lukosius joined the Navy to see the world, and what did he see? He saw the sea! For the first three days on board the *Pillsbury*, Lukosius has the worst case of seasickness. After that, he never gets seasick again.

The *Pillsbury* is a feisty ship, smaller and slower than a regular destroyer and designed to escort merchant marine convoys. With a tighter turning radius, she is built for hunting and killing German submarines.

Mack Pickels keeps a war diary. His adventures on the *Pillsbury* take the young Texan to Valencia, Spain, and to the exotic city of Casablanca. South of the Azores one night, he is listening to music on a German radio broadcast from Berlin when a news bulletin interrupts. The US Navy has lost five aircraft carriers in the Gilbert and Marshall Islands campaign in the Pacific. Mack recognizes the report for what it is: total bullshit.

"We had a big laugh over that," he writes in his diary that night. Another claim from Berlin radio he also finds funny: accounts of "nuisance raids over Germany." Yeah, right. Some nuisance. That must mean the Germans are getting pounded.

■ ■ ■

Far from the USS *Pillsbury* and the Mediterranean, in the island nation of Iceland near the Arctic Circle, Commander Daniel Gallery is making the best of his frustrating existence.

He has seen to it that the US Navy base is now a thriving little village with gravel roads and comfortable living quarters. In another display of his quirky sense of humor, he has his chief metalsmith manufacture two palm trees, which he plants at the entrance to the base. Even the austere chief of naval operations, Admiral Ernest "Ernie" King—aware that no trees grow in barren Iceland—has a good laugh over that one when he inspects the base. Gallery authorizes a new certificate suitable for framing for any man serving one hundred days in Iceland—an "FBI," as in "Forgotten Bastard of Iceland."

He feels like a landlubber. He knows he's taking part in a consequential mission in the Battle of the Atlantic, but the real glory is taking place in the Pacific, with the historic battles of Midway and Coral Sea and Guadalcanal. Reading newspaper accounts and confidential action reports from the Navy about these great events drives him to distraction and, he must admit, episodes of jealousy.

At last, in May 1943, his prayers are answered. Gallery is ordered home to take command of a fighting ship.

Hallelujah! Gallery bids farewell to the "Godforsaken hole" of Iceland and flies to Washington for a few weeks of rest and relaxation and quality time with his wife, Vee, and their three kids at the family farm in Virginia. One month later, when his leave is over, he is off again, but this time to Astoria, Oregon, and it is there that he takes command of the aircraft carrier USS *Guadalcanal*.

FIVE

SIN CITY

U*-505* is on a roll.

The day after torpedoing the *West Irmo*, the *U-505* lookouts spot another plume of smoke. It is 1:07 p.m., April 4, 1942, about 160 miles off the Ivory Coast. The plume is far to the south. Loewe figures the vessel—whatever it is—has a twenty-five nautical mile head start. The chase begins. With diesel engines running at top speed, it takes Loewe almost seven hours to catch up to the ship. He remains on the surface as he approaches, steering straight at his target to present the thinnest possible silhouette.

The crew hears the intercom click. It is Loewe.

Auf gefechtsstationen! "Battle stations!"

U-505 has not been spotted yet. The target has no escort. At eight hundred meters, Loewe lets loose with a single torpedo.

It's a hit, slightly astern.

For the next few minutes, Loewe and his men watch as the merchant seamen on the ship they've just blown up scramble into lifeboats. Her bow is steadily rising and, for a moment or two, hangs high in the night air. Then, just like that, she's gone.

Loewe is almost casual as he orders *U-505* to pull up to the four life-boats now bunched together in the ocean. Loewe figures the enemy ship has gone down so rapidly the radio operator had no time to send out an SOS.

She's the Dutch merchant ship *Alphacca*, and Loewe is wistful when he discovers she's out of Rotterdam. Loewe's mother is of Dutch descent. She taught him as a child to prefer tea over coffee. Loewe had even shown the German cook on *U-505* how to brew a decent pot of Dutch tea, which on the cook's first attempt tasted so bitter it was undrinkable. In a sense, these merchant seamen he has just shipwrecked are his kinsmen. He starts speaking to them in English, but then swiftly shifts to German.

The ship's master is Reindert Johannes van der Laan. At fifty-nine, he's old enough to be Loewe's father. The ship is out of Cape Town, on her way to Freetown, carrying two thousand tons of copper, four hundred tons of zinc, and twenty-five tons of vanadium, an essential war metal in the manufacture of tools, pistons, and axles. An interesting cargo to say the least.

Loewe takes special steps to see to it that the Dutch seamen have enough supplies and water. He even wishes the fifty-two survivors good luck and bon voyage as they make their way to the Ivory Coast. Of course, nothing can be done about the fifteen men he has just dispatched to the bottom of the sea, but war is war.

It's time to go. He orders *U-505* to proceed on a southerly course—but it's a feint. He wants the survivors on the lifeboats to think that's his route when they are picked up and questioned by Allied intelligence about the bastard U-boat that sunk the *Alphacca*. Once he is far from their sight, Loewe changes course to his intended track, which is northerly.

That night, in his cabin, Loewe writes in his war diary: "Irony of fate, we fight against people who speak our language."

It's as if sinking a ship full of his kinsmen triggers a world-class run of bad luck for Loewe. For the next two weeks, the ocean is empty—not so much as a fishing boat is sighted. After that, every ship Loewe chases down turns out to be neutral or a British warship, which he is under orders to give a wide berth. The threat from the air is relentless—British bombers are a constant presence. Crash dives become a daily occurrence. The crew is exhausted and showing signs of scurvy—nasty skin eruptions and loose teeth—due to the depletion of the boat's fresh fruit supply. The diesel engines need an overhaul. And most worrisome of all, *U-505* is in danger of running low on fuel. Nerves are getting frayed, and it's showing with glum faces and restless nights trying to get some shut-eye. Machinist Hans Decker, for one, is sick of seeing the "same faces day in and day out, listening to the same stories."

Even gung-ho Hans Goebeler, who dreams of returning to his village in Germany with a chest full of medals, agrees it is "clearly time to head home."

On April 27, Loewe clicks on the boat's intercom to make the long-awaited announcement.

"Beginning today, we start our trip home."

The next morning, with the change in course, the sun now comes up on the starboard side of *U-505*. The boat threads her way past the Azores and up the Spanish coast. As they get closer to the Bay of Biscay, razors and scissors appear and so commences the grooming of bushy beards and long hair. The men haven't had a real shower in almost three months. Fresh water on the submarine is a scarce commodity. There is an apparatus that turns seawater into potable water, but the quantity is limited and saved for meals and brushing teeth. For daily rinsing of the body, there is rubbing alcohol or seawater with a special soap that lathers up despite the briny salt. Loewe himself doesn't have much sprucing up to do. During the entire voyage, he's been finicky about personal hygiene and has stuck to a regular schedule, shaving every fourth day as an example to his men of military discipline.

The morning of May 7, after eighty-four days at sea, *U-505* enters the port of Lorient in occupied France. The crew assembles on the deck in formal dress. Four pennants flutter in the wind, signifying four kills. Up ahead they see dignitaries and spectators on the pier, waving and hollering. Nice, but most of the crew's attention goes to some very pretty nurses, who sure seem happy about the arrival of this woebegone and hairy crew.

"Three cheers for *U-505*! Hooray! Hooray! Hooray!"

The men of *U-505* disembark. After all this time at sea, they're a little wobbly. It'll take a while to shake off their sea legs and walk steadily on land. How wonderful it is to be out of that steel cigar, breathing fresh air!

They're dying for a shower, but their first stop is the dining room hall for mail call. Giant bags are opened, and as the postman calls out the names of the sailors they spring forward to collect their letters and parcels from home. Christmas has come early.

Hans Goebeler sits quietly in the corner and with downed eyes reads the bundle of correspondence from his parents in Bottendorf. His mother is a devout Christian. Each letter opens with a verse from the Bible and ends with a reminder that he should pray as she had taught him when he was a little boy.

For some of Goebeler's comrades, the news from home is life altering. A baby has come into the world. A brother has been slain fighting the Russians on the Eastern Front. A cherished aunt has passed away. Our house has been bombed by the British and the family has moved to the countryside for safety. Stay brave, my son. Make us proud. Victory will be ours.

Tears are wiped away as the letters are carefully folded like cherished family mementos and tucked into their pockets. Who knows when or if they'll ever see home again?

At last, it's time for long, hot showers. What luxury, compared to those quick seawater hose-downs, or thunder showers on the top deck during tropical rainstorms. Nothing like this! Hot water never felt so glorious.

The banquet that night is not to be forgotten. Juicy sausages. Thick white bread. *Fresh vegetables.* Mouthwatering French desserts. And real beer—Becks and Falstaff—not the crappy brands they stored on the boat.

Toast after toast is made. Some of the boys get sick from gorging on rich food and alcohol, and these unfortunate souls spend the next few hours throwing up. The rest, their bellies full, sit back and savor cigarettes and pipes, enjoying the space, and the sheer joy of this respite from life on *U-505*.

The events of the day leave them exhausted, but after a night of deep rest in their barracks, they get right down to business. Half the crew is sent on furlough for a full week. When they return, the other half will depart for home. Goebeler decides his village is too distant to make it there and back in a week. Plus, the train connections in time of war are not reliable. He hopes he can visit his family on his next leave.

U-505 is hoisted onto dry dock for routine repairs. The engine oil is changed, the equipment tested. Aerial attacks by the British are met with a dismissive shrug since the bunkers are impenetrable with reinforced concrete.

Goebeler works hard, but his nights are dedicated to pleasurable pursuits. There is a German club for homesick boys where they can watch the latest Nazi propaganda movies and drink cheap beer or play cards and just relax. They even listen to "decadent" American jazz.

That's nice, but the real lure of Lorient is the so-called sin strip. Goebeler strolls it wide-eyed, hearing the intoxicating rhythm of French love songs serenading him, luring him. Oh, those French mademoiselles in fancy dresses with flowers in their hair, the heady scent of perfume in the warm night air. The lipstick! The rouge! No German girl Goebeler ever knew would dare walk the streets of Bottendorf like this. Should a sailor prefer a pretty fraulein who speaks the mother tongue, German prostitutes are available. Kriegsmarine doctors regularly inspect the ladies for venereal disease. Kapitänleutnant Loewe prefers not to even think about the "bad influences of the seaport."

Goebeler has never been with a woman before. He can talk a good game, but all he knows of the opposite sex comes from watching military training films.

In time, he takes a special interest in a pretty French girl who calls herself Jeanette. She has soft, light brown hair and a fetching, slender figure. Goebeler is drawn to her from the moment he encounters her. They even go out on traditional boy-girl dates—to a French bistro for dinner and a salon where Jeanette has her hair styled. Strolling down the street, she pops into a gift store and buys Goebeler a pendant of Saint Christopher, the patron saint of travelers.

"I want you to have it. He will bring you safely back to me," she tells him.

To Goebeler's inexperienced eye for jewelry, it looks like solid gold.

A deep affection develops between them. Goebeler even wonders if there might be a life together, after the Nazis win the war.

It never occurs to him that she might be working for the French Resistance. No suspicion whatsoever. The boy is blinded by lust.

The month passes quickly. By Sunday, June 7, 1942, this idyllic time of rest and relaxation is over. The entire crew has reported back to duty from their furloughs, and *U-505* is ready to ship out again on its second wartime patrol.

"The Great Lion" himself—Admiral Doenitz—is in Lorient on an inspection tour and bids the crew farewell. He strides across the gang plank as the men stand at attention and engages in brief words with Axel-Olaf Loewe. Then he writes this into the boat's official war diary:

> *First mission of captain with new boat, well and thoughtfully carried out. Despite long time in operations area, lack of traffic did not permit greater success.*

Loewe accepts it for what it is: a passing grade, but no gold star. It's an accurate assessment of *U-505*'s middling war record to date, and an exhortation that he and the crew do better.

They will certainly try, and they'll need all the luck they can get.

Sailors are notoriously superstitious. Loewe, for instance, wears a penny talisman on a chain around his neck—a gift from his wife—and is abiding by the custom that no U-boat skipper ever depart on a Friday, for if you do the entire crew will think the mission is doomed. You can depart at 2359 on a Thursday or 0001 on Saturday, but never, never, never on a Friday.

The military band of the Second U-boat Flotilla lines the submarine pen and plays stirring music as *U-505* slips out to sea. Once again, Loewe orders the garlands of ceremonial flowers that adorn the conning tower be tossed into the Bay of Biscay while land is still in sight—another superstition he and the crew are unwilling to challenge.

Under cover of darkness, Loewe runs the surface of Biscay at top speed. At dawn, Loewe dives to avoid British air patrols. They can hear distant bombs and depth charges dropped by the British, who are in hot pursuit. Time to open his sealed orders and read them to the crew:

> "Men, we've been ordered to the Caribbean, and so we have a long run across the Atlantic to make. I don't have to tell you, the bridge will have to keep an especially sharp lookout. We will be crossing many commerce routes. In fact, the success of our trip will depend

on how well our bridge watch does its job. The anti-submarine techniques of our enemy are improving day by day. Therefore I require a continuous improvement of my crew. *Das ist alles.*"

The men are pleased. The Caribbean! Fertile hunting grounds for American freighters coming through the Panama Canal. Many U-boats assigned to war patrol in the Caribbean have made triumphant returns to Lorient, with their torpedo bays empty and six or more victory pennants flapping in the wind.

When *U-505* surfaces, the sea is serene. Six days pass. All is good. *U-505* approaches the Azores and then veers westerly across the Atlantic. Nothing but blue water and pods of porpoises, schools of flying fish, and the occasional albatross with its enormous wingspan. They are making good time. They've never sailed this far with so few attacks from the air. The emboldened crew even sets up a mess table on the top deck where they enjoy lunch alfresco.

Hans Goebeler is on bridge watch duty as *U-505* gently dips and sways in the ocean. The night sky blazes with galaxies. Three weeks of blissful peace. At midnight, the cook, Anton "Toni" Kern, appears with a pot of *Mittelwachter*, strong black coffee laced with rum. These days are the best.

Then, on the twenty-seventh day out, they see a black smudge on the horizon.

■ ■ ■

The USS *Sea Thrush* has been known by several names and under several owners during its long existence. When the steamer was launched in 1920, she was called the *Delanson*. Nine years later, she was renamed *Exilona*. Now, under the ownership of the Shepard Steamship Company out of Boston, she is called *Sea Thrush*.

By any name, she is one tough vessel.

It is 11:55 a.m., and her master, Arthur Hunt, doesn't realize that his ship has only seven hours left to exist.

Hunt is carrying ammo and other war goods in the cargo hold. He left Philadelphia a week earlier and is en route to Cape Town for refueling, with a final stop in the Iranian seaport of Bandar Shahpur. The ultimate destination is the Soviet Union. Twenty twin-engine airplanes are lashed on the deck. On board are sixty-six merchant marines, armed guards, Army officers, and technicians.

And she's a turkey dinner of a target.

Sea Thrush is 425 miles northeast of San Juan, Puerto Rico, when it happens. No one aboard *Sea Thrush* knows at the time that they have been stalked for the last seven hours by a devil under the sea, *U-505*, and that they are now only seconds away from impact. Two torpedoes, racing at them at a depth of three meters, are on the way . . .

Boom!

The first eel hits port side forward of the collision bulkhead.

A heartbeat passes.

Boom! A second blast, just behind the bridge.

Gun crews man battle stations at the main gun deck, braced to open fire with .20mm and .30mm antiaircraft guns. They may be going down, but not without a fight.

As the *Sea Thrush* slowly sinks, the Americans wait for the damn U-boat to pop out of the water. *Come on, come on . . .* They have a big surprise in store for the Nazis who just torpedoed them.

But an hour passes, and still no U-boat. How long can they wait? The sea is about to wash up over the guns.

That does it. The *Sea Thrush* is toast. Hunt reluctantly orders his crew to abandon ship, and the men fill the lifeboats.

But *U-505* isn't done with her yet. The Stars and Stripes are still proudly flying on the abandoned *Sea Thrush* when Loewe delivers the knockout punch, a third torpedo on the starboard side. *Sea Thrush* breaks in two and sinks to the bottom of the sea.

By some miracle, there are no casualties. Hunt's freighter may have been obliterated to kingdom come, but it could have been worse. Hunt must wonder if the U-boat skipper waited for his men to lower the lifeboats before firing the coup de grâce.

An act of mercy, in the middle of a war?

SIX

SITTING DUCK

B ack in his glory days as a champion all-American swimmer, Albert Rust Jr. never dreamed that his most important swim would be on the open ocean after his ship had been torpedoed by Nazis.

Then, on an ordinary workday as an office manager at a Great Lakes steel company, Rust's boss delivers some startling news:

"Your number's up, Al. You're next."

He's talking about the draft. *Uncle Sam wants you!* At this point, Rust is twenty-four years old and can't see himself as a soldier—so this Zionsville, Indiana, native enlists in the Navy, thinking he's made the right choice.

But not for long. They assign him to the Navy Armed Guard, which is about the most dreaded assignment any sailor can get.

Four weeks of training on a "little boat" in Lake Michigan, and that's pretty much it for instructions. The next thing he knows, Rust is put in charge of a ten-man gunnery crew. The Navy orders them to board a train for New York City, and at the Brooklyn Navy Yard, they are assigned to a merchant ship, the USS *Thomas McKean*. This is to be the ship's maiden voyage.

The *Thomas McKean* is carrying ammo, gasoline, lumber, pipe, Jeeps, and six twin-engine planes latched to the deck. Her route is the same as the *Sea Thrush*. After refueling in Trinidad, she is to round the Cape of Good Hope, then move on to the Iranian port of Bandar Shahpur.

Ultimately, the war material is to be transported by land to the Soviet Union.

Rust is plenty upset. He can't figure out how he got appointed gun captain. He is given a grand total of forty minutes to teach the Navy Armed Guard unit on the *Thomas McKean* how to fire the .20mm gun. They are kids—sixteen, seventeen years old. Rust is the "old man" in the unit. Maybe that's why they put him in charge. That, or maybe his broad shoulders from all that swimming.

What a joke. Rust feels like a sitting duck. Merchant ships are sluggish and clumsy vessels. Due to the cargo they carry, they are also priority targets for German U-boats. The *Thomas McKean* is sailing without a Navy destroyer escort. Nor is she joining a convoy. Good luck. If a Nazi submarine spots them, there's nothing the freighter can do. They all know it. The U-boat will just keep hitting her with torpedoes until she sinks.

And this is where Rust finds himself on April 28, 1942, one day after the destruction of *Sea Thrush*. The *Thomas McKean* is zigzagging at ten knots. Rust is in his bunk. It's a hot night in the Caribbean Sea, and he's in the nude, sweating, trying to get some shut-eye in the sticky, tropical heat, when a tremendous blast shakes the freighter. (If you haven't already, read the Prologue for more detail.)

Rust leaps out of the top bunk and almost lands on the head of the kid in the bunk below. The swimming trunks he'd been wearing the night before are draped around a rail, and Rust hurriedly slips them on. That's all he's wearing. No helmet. No life jacket. He does manage to grab his wallet. Then he flies down the ladder and finds himself on the main deck. What he sees is shocking.

There's a giant hole at #5 hold. The stern gun has been blown up. Three of his guys from the Armed Guard unit are dead. A merchant marine is lying on the deck in a pool of blood and water.

Rust runs to his assigned lifeboat, #4. The lifeboat is already suspended about ten feet over the side of the ship, so Rust has to lean over and grab hold of the davit. Big mistake. He burns his hands going down, loses his balance, and lands on his back. On the lifeboat he sees Jack Hannon, his buddy from Ohio. Hannon's left hand looks as if a knife has splayed it in half. It's bleeding badly, and he is in severe medical shock.

They wash the wound out with salt water. It stings like hell, and they wrap the hand in bandages as best they can.

Twenty minutes or so later, a submarine surfaces. It reminds Rust of the great white whale Moby-Dick—the U-boat hull has itself turned white from briny seawater. German sailors are on deck, grinning with pleasure at the sight of the American freighter they've just torpedoed.

Thomas McKean stubbornly refuses to sink. Apparently, the U-boat skipper doesn't want to expend another torpedo to do her in. Instead, he orders gunners to fire relentlessly at the *Thomas McKean*. She absorbs nearly a hundred rounds from the U-boat's .105mm guns before at last rolling over, peppered with holes.

Once the U-boat takes off, the Americans don't know what direction to go. Lifeboat #4 is equipped with a Boy Scout compass that doesn't really function in a metal boat. Which way is west? Not even the seasoned merchant marines stuck in the boat with Rust can figure it out. As the saying goes, it's FUBAR, Fucked Up Beyond All Repair.

The men on Lifeboat #4 hoist a red sail and yellow flag to signal distress. Puerto Rico is about four hundred miles away.

Over at Lifeboat #3, Rust sees the body of a merchant marine named Russell Funk being put overboard and buried at sea. The body doesn't want to leave. It just stays floating alongside the lifeboat until somebody gives it a good shove with an oar and it drifts away. It's a sight Rust is pretty sure he'll never forget.

Along with Rust, there are fourteen men in the twenty-two-foot-long Lifeboat #4. The senior officer in charge is Third Mate William Muci, and he's got his work cut out for him. Because the compass doesn't work, he'll have to navigate by the stars. Muci starts by rationing the water— four ounces a day per man. There's not much food—some Horlicks malted milk tablets, chocolate bars, and pemmican beef jerky with nuts and fruit—an excellent source of dextrose. Unfortunately, the lifeboat isn't equipped with fishing tackle.

Night falls and Rust starts to shiver. All he's wearing is his swimsuit. Not even shoes. A blanket offers some comfort. There's a lantern but the decision is made to proceed without lights in case a German U-boat decides to take a potshot at them.

The cook, Albert Brooks, is certain they're going in the wrong direction.

"Oh, Lord," he mutters to no one in particular. "I know they're wrong."

What lousy luck. Rust can't believe it. He signs up for the Navy on February 18 and is sunk on June 29.

For the first two days and nights, all four *Thomas McKean* lifeboats manage to stay within sight of one another. They're spread about a mile apart. By the third day, the whims of wind and sea pull them in different directions. They're on their own.

Rust is surprised to discover that he is having bladder and bowel issues in the lifeboat. He finds it challenging to take a leak or a dump off the side because the lifeboat is continually bobbing up and down. The motion shuts down the body's sphincter. They all end up passing a bucket.

On the third day out, a Navy PBY Catalina amphibious plane spots them from the air. After circling the lifeboat, a seabag drops out of the plane. What a heavenly sight—and it's Rust's turn to show what he can do.

"I'll go out and get it," Rust says. The former all-American swimmer knows he can handle the job.

They tie a rope around him and Rust dives in. He's still got the strokes that won him medals for the 220-yard freestyle events at the 1936 and 1937 National Swimming Championships. He manages to grab the seabag, swim back, and haul it on board to cheers from his fellow sailors. The bag is a lifesaver, loaded with canned food.

Before flying off, the PBY signals the lifeboat to continue sailing west. Figuring they're about to be rescued any hour now, Rust and the other men have a "feast," such as it is. They drink all the remaining fresh water and wait to be picked up. And wait. Finally, it dawns on them: There will be no rescue vessel. The only source of water now is whatever rain they can catch in the sail and blanket.

Three more wretched days pass. They're parched and feeling delirious. Their lips are splitting. Is this the end?

No. They reckon they *must* be getting close to land because they can smell earth. They can't see it, but they can smell it. And something delicious. Maybe chopped sugarcane.

Somebody shouts, "I smell green."

It's the Dominican Republic, a welcome smudge of green on the horizon. Lacking working oars, Rust and the others start using their feet to row. They can see a Dominican on the beach waving his arms. Then a tiny rowboat comes out to greet them. The guy on the rowboat doesn't speak much English, but enough to communicate to the Americans they'd better get their feet out of the water pronto because the barracudas and sharks might find them tasty.

Rust has lost thirty pounds in the ordeal.

When they finally get to the capital of the Dominican Republic and speak to a diplomat from the America legation, Rust learns the fate of the other three *Thomas McKean* lifeboats. Turns out, his Lifeboat #4 fared the worst of it. They were at sea the longest—nine days—the consequence of that malfunctioning compass.

Not that the other men had a picnic.

Lifeboat #3—commanded by Second Mate Roland Foster—is the only boat equipped with a motor, which means no sail, and he has to ration the gas. Foster jerry-rigs a jib using the distress flag. He's worried about running out of gasoline. What if he hits a squall? If the wind blows away his jib, where will that leave him?

He has no navigation charts although he does have a sextant and he is a skilled navigator.

One thing that really gets his goat is the keg of water on board Lifeboat #3. It's sour, which is disgraceful considering this is the *Thomas McKean*'s maiden voyage. He and the men drink the water anyway because they have no choice, but it tastes nauseating. They struggle to keep it down.

After four days in the lifeboat, a PBY bomber spots them and drops a load of food, including pemmican and malted milk. The plane circles for quite some time before flying southwest and out of sight.

On the Fourth of July—six days after the sinking of *Thomas McKean*—Foster gets a good latitude reading with the sextant and figures he's sixty miles north of the British West Indies. He starts the motor and runs it until dark. The next morning, he sights land and restarts the motor, landing at the northernmost tip of Anguilla.

All he's got with him are the clothes on his back, his second mate's license, and the sextant. And a really nasty sunburn.

Lifeboat #1 fares best of all, probably because it has the ablest seamen—Master Mellin Respess and Chief Engineer Thomas McCarthy.

On day three, at 9:00 a.m., they are sighted by an American plane that comes swooping down. A box is dropped out of the plane. Inside is a message. Respess reads it out loud.

Hold steady. Keep to the course they are currently steering, south by southwest.

Nine hours later, a minesweeper appears on the horizon and comes alongside Lifeboat #1. Respess, McCarthy, and the other sailors and merchant marines climb aboard, are issued warm clothes, and eat a light meal of soup.

Then they are taken to St. Thomas in the US Virgin Islands where they are reunited with the men from Lifeboat #2, which had landed on the island of Saint Kitts in the West Indies.

McCarthy is still plenty pissed when he finally gets to shore in Norfolk, Virginia. Old salt that he is, he just has to vent.

"I lost everything I had on board. I had a brand-new outfit, and I didn't get anything off of the ship except what I had on my back. I had no shoes on even." He writes a list of what he's lost and makes it clear he's expecting full reimbursement.

Twenty days after his rescue, Respess was returning to the United States on a merchant ship, the SS *Onondaga*. Five miles off the coast of Cuba, *Onondaga* was hit by a single torpedo from *U-129*, and sank within a minute. All hands jumped overboard. Nineteen merchant marines drowned, including Mellin Respess. He was forty-one.

It would be years before the survivors of the *Thomas McKean* learned the name of the U-boat that sank their ship.

SEVEN

RELIEVED OF

COMMAND

F ollowing the sinking of the *Thomas McKean*, *U-505* stalks the waters
off the coast of Colombia for the next sixteen days in search of prey,
but the sea is devoid of enemy ships. Annoyed by the lack of action,
Kapitänleutnant Axel-Olaf Loewe sails due south to the coast of
Venezuela. On July 27, 1942, the lookout on the bridge of *U-505* catches
sight of a magnificent three-masted schooner. Curiously, there's a luxury
automobile lashed to the deck.

A luxury car on a ship's deck? Very strange.

Loewe orders battle stations. In less than thirty seconds, the .105mm
cannon is loaded and ready to fire. Loewe stares into his binoculars and
analyzes the state of play. The schooner is flying no flag. She's zigzagging
at an aggressive clip. Definitely suspicious. Loewe wonders if the schooner is bait. Is *U-505* being lured into a trap?

"Fire a shot over her bow so we can find out who she is," Loewe tells
gunnery officer Gottfried Stolzenburg.

It's meant to serve as a warning shot, but the cannon ball falls short of
its intended target—the open sea—and demolishes the schooner's main-
mast. Did Stolzenburg misunderstand the order? Or was it an unfortunate accident of war?

Loewe can see the passengers and crew abandoning ship. He can also make out a name painted on the side of the vessel—she's the *Roamar*, out of Cartagena. Colombia is an important strategic ally of the United States due to its proximity to the Panama Canal, although the nation is officially neutral in the war. As of now.

Uh-oh. What to do next? Loewe struggles with the answer. He fears he has blundered his way into a diplomatic incident. He has seconds to make up his mind. When he does, he comes to the conclusion that the only course of action is to cover up any evidence that the *Roamar* existed.

"Sink that thing, but quick, Stolzenburg."

Two more shots from the cannon finish her off, and the *Roamar* is splintered into bits of flotsam. Loewe stares out at the sea. There are no signs of survivors. The schooner is gone.

U-505 moves on, but Loewe can't shake the feeling that he has made a calamitous misjudgment.

As the days proceed, Hans Goebeler observes his beloved skipper. Loewe is showing symptoms of "nervous distress." His skin grows paler by the day. Loewe rarely leaves his cabin. Very soon, the crew comes to realize these symptoms can't be attributed just to anxiety. Medically, something is very wrong.

Herbert Nollau, the first watch officer and a 1936 graduate of the German naval academy, assumes command of *U-505*. He radios U-boat headquarters informing Admiral Doenitz that Kapitänleutnant Loewe is seriously ill. The response is swift: *U-505*'s mission is declared terminated, and she is ordered to return to Lorient.

When Loewe is told, he isn't surprised. In German naval tradition, nothing can be accomplished unless the commandant is in fit condition.

U-505 turns east for Europe and her home port of Lorient, but she is running very low on fuel. A rendezvous with a "*milch* cow" is arranged. The milk cow is a U-boat modified to carry extra tanks of gasoline, provisions, and torpedoes. Five hundred miles south of Bermuda, *U-505* links up with *U-463* and via a tether pipe sucks twelve tons of diesel fuel.

The final leg of the journey is harrowing. *U-505* runs the entire stretch of the Bay of Biscay underwater due to the pervasive presence of British air patrols. She surfaces only to recharge her batteries and take in fresh air, then submerges before she is discovered.

On August 25, 1942, *U-505* finally sights Lorient Harbor.

As before, a military band and German dignitaries stand on the dock in greeting. The men of *U-505* assemble on deck in their dress uniforms. Hans Goebeler's heart leaps. There's Jeanette, with quite a few other lovely French ladies.

Goebeler is delighted by the presence of his girlfriend. But something is different this time. The homecoming seems to him "a little hollow." Is it because their mission has been aborted and they have returned with just two kills? He understands the situation better once he steps off the boat and realizes the seaport has been the target of an intense Allied bombing campaign. Although the thick concrete U-boat pens are impenetrable, the port of Lorient is under siege. Much of the city lies in ruins.

Axel-Olaf Loewe struggles to make his way off the boat. Very quickly, German naval doctors diagnose chronic appendicitis. An emergency operation takes place two days later. As for the sinking of the *Roamar*, Loewe is correct to fret. The schooner is owned by a prominent Colombian diplomat. Colombia is now threatening to declare a state of war against Nazi Germany. Loewe's reputation is sunk—just like the *Roamar*.

He is relieved of command.

■ ■ ■

The new skipper of *U-505* is Kapitänleutnant Peter Zschech. Just twenty-four years old, he is Germany's youngest U-boat skipper, nine years younger than the man he is replacing.

Loewe is taken aback when he learns about the appointment. He knows Zschech. Genetically, Zschech (pronounced "Check") is a "racially pure" Aryan, but in Loewe's opinion he has a "rather weak constitution." Loewe wonders whether his successor is really cut out to command a U-boat, which "uses up a man faster than anything."

Hans Goebeler takes stock of Zschech soon after he boards *U-505*. Intelligent. Cultured. Confident. Also, arrogant. Moody. Aloof. Imperious.

Drinks too much. Worst of all is his explosive temper, so opposite to Loewe's fatherly presence.

Every U-boat skipper is entitled to his own emblem, and Zschech wastes no time scraping off his predecessor's axe-wielding lion and replacing it with the Olympic Rings, the emblem for the year he graduated from the naval academy, when Berlin hosted the 1936 Olympics.

Joining Zschech are two senior officers. Óberleutnant Thilo Bode is the new executive officer, replacing Herbert Nollau, who has been promoted to captain of his own U-boat, *U-534*. Thilo Bode graduated from the same class as Zschech. The nature of their relationship is a puzzlement. They spend long hours huddled together in Zschech's cabin. Goebeler claims he sometimes sees them "hold hands."

"A thoroughly unpleasant character," is Goebeler's assessment of Thilo Bode.

The chief engineer and third in command is Óberleutnant Josef Hauser, who is cut from the same cloth as Zschech when it comes to arrogance. The crew quickly brands Hauser "The Raccoon." Why Raccoon? Because Hauser is always preening, grooming his beard, and making sure his hair is just so. He never met a mirror he didn't like, and sometimes the crew catches him posing and scowling, like Mussolini. He also walks with a swagger and thinks he knows everything about engines when in reality—at least in Goebeler's opinion—he knows "almost nothing." Even Thilo Bode agrees that Hauser is "not an easy man" and wonders whether Hauser has the qualifications and experience to be chief engineer. Hauser is only twenty-two. He joined the Nazi Party when he was an eighteen-year-old student.

U-505 undergoes repairs and modifications and is loaded with fresh supplies of torpedoes and provisions. Mighty tanks are installed, ones that considerably boost the fuel supply and permit *U-505* to stretch its time at sea on war patrols. *U-505* is also outfitted with a new contraption—an early warning radar apparatus called a Metox, which looks like a crude cross-shaped antenna and can detect enemy aircraft when *U-505* is riding the surface, long before the lookouts can eyeball them in the sky.

On October 4, 1942, after five weeks in port, *U-505* is ready to depart on her third mission. The chief of Western operations for the Nazi

U-boat fleet writes a farewell message in the boat's log. When he's finished, the crew gathers around to read:

The sinking of the Colombian sailing ship would have been better left undone. No further remarks.

It's a stinging rebuke, and a hell of a farewell note.

At 6:00 p.m., under cover of approaching darkness, *U-505* sets off again. All nonessential personnel assemble on the deck and wave at the cheering crowd.

Not one hour later, Second Watch Officer Gottfried Stolzenburg orders garlands of flowers decorating the conning tower thrown overboard, the usual tradition to fend off bad luck. Nobody has ever defied this tradition, until now. Zschech screams at Stolzenburg to halt.

Stolzenburg can't believe it. He tries to explain that the flowers must be tossed over before losing sight of land, but Zschech cuts him off.

"Kapitänleutnant Loewe is no longer in command of this boat! This is my boat and I am the only one giving orders from now on. I want everybody to understand that!"

And so the flowers stay, to the dismay of the crew. Why tempt fate? It makes the crew wonder about Zschech's wisdom.

Two days out, *U-505*'s receives its orders via radio. The destination is Sea Square ED99 under the German Navy's secret code that divides the oceans into quadrants rather than longitude and latitude. This translates into the Caribbean Sea—specifically, the coastal waters off Trinidad. The target: tankers emerging from the mouth of the Orinoco River, carrying oil from Venezuela.

It takes *U-505* two more weeks to traverse the Bay of Biscay. At last, they are well on their way across the Atlantic. But questions about the competence of the new skipper and his senior staff consume the men.

And it doesn't take long for the flower fiasco to bring bad luck to *U-505*.

Josef Hauser, a.k.a. Raccoon, is supervising a test crash-dive. Diving is a tricky procedure on a submarine. It involves turning several dozen hand wheels and water valves in a perfectly timed sequence until the tanks fill with the proper tonnage of water. One mistake could doom the boat to oblivion.

Hauser opens the bow diving tanks but is too slow opening the aft tanks. There is an immediate imbalance of weight that sends the submarine into a death dive. In seconds, *U-505*'s nose is at a 42-degree angle and comes close to hitting the seafloor. Hauser could have killed everyone on board. Engineer Fritz Foster takes over and blows the tanks. The submarine levels out. When he's certain the danger has passed, he turns on the Raccoon.

"The submarine is not your toy! There are forty-nine human beings on this boat and they want to return to their homes out of this patrol!"

Foster's justified rant has no visible effect on Hauser.

Meanwhile, Zschech and Thilo Bode seem to take sadistic pleasure in treating the crew like dirt.

Hans Goebeler is in his bunk one night off the coast of Trinidad, dead asleep. He's got six hours to go before his next shift starts when he's shaken awake. He's informed that Óberleutnant Bode requires his immediate presence on the bridge. Assuming it's an emergency, Goebeler takes less than sixty seconds to dress and report to the bridge. He snaps to attention.

"Maschinengefreiter Goebeler here as ordered, sir!"

"Goebeler, get us some coffee, and hurry!"

Goebeler can't believe it. Thilo Bode has awakened him to pour a cup of coffee? Seriously? It's unthinkable. Goebeler hustles down the ladder to the ship's galley. His buddy, the cook Anton "Toni" Kern, is on duty and there's already water boiling on the burner. When the coffee is ready, Goebeler climbs the ladder, carrying a pot and three cups.

All seems well. Goebeler can now return to his bunk. As Goebeler makes his way down the ladder, he suddenly cries out at the searing pain on his scalp. Bode, that miserable bastard, is pouring his entire cup of hot coffee right over Goebeler's head! And he's screaming at Goebeler.

"You *idiot*! I said fresh coffee, not this stinking bilge water. Get down there and make me some real coffee immediately!"

All Goebeler can do is clench his teeth. In the tyrannical traditions of the German Navy, it is unthinkable for him to show any sign of disobedience, no matter how justified. He'd be facing severe punishment or even court martial.

"*Jawohl, Herr Óberleutnant!*"

Toni heats a fresh pot. "Here," the cook tells him. "Give that asshole his cup of coffee. This is what you must do, Hans. Fill your mouth with coffee until you're on the last couple of steps of the ladder. Then spit enough coffee back into the cup until it is filled right up to the top. You'll see how he likes that cup of 'fresh' coffee."

Goebeler does just what the cook suggests and watches as Bode takes a sip and savors the fresh brew. *Drink up, arschloch!*

Goebeler can't help smiling as he returns to his bunk. He may have lost an hour of sleep but he figures that Bode got what he deserves.

Later, on the night of November 7, 1942, Goebeler is getting some shut-eye when he is awakened by the trembling of his bunk and the wild jerking of the submarine. He's up with a jolt. The boat's diesel engines are running at top speed. Goebeler doesn't have to wait for the call to battle stations. The roar of the diesels tells him the chase is on.

Zschech is positioned on the conning tower. Through his binoculars, he sees what looks like a seven-thousand-ton freighter. A nice prize, if he can get close enough to fire off his torpedoes. But she's a speedy vessel and despite a frantic two-hour chase, *U-505* can't seem to catch up to the enemy ship. In frustration, Zschech decides to open fire from fifteen hundred meters.

Torpedo #1—fire!

Torpedo #2—fire!

The seconds tick by. Nothing. It's a double miss.

Zschech's face turns red. He utters a curse.

Two thousand meters now separate the freighter from *U-505*. Zschech unleashes torpedoes #4 and #5. It's a long shot. The maximum distance for a torpedo is four thousand meters. Alex-Olaf Loewe never fired from a distance greater than fifteen hundred meters because he knew he'd probably be wasting a precious eel.

A full minute ticks by. Then two minutes. The torpedoes must be spent. Zschech is about to turn away in disgust when he hears an explosion, followed four seconds later by another blast. Two direct hits! In the blackness, far away, they can see lifeboats being loaded. The Germans are ecstatic. They've never experienced a successful torpedo fired from this range.

Zschech is very pleased with himself and orders *U-505* to make a hasty exit. *What?* Hans Goebeler for one is taken aback. Common decency and the laws of the sea require Zschech to check on the condition of the survivors and render assistance, if he was able. At least, that was how it worked when Axel-Olaf Loewe was skipper. It is the fellowship of the sea that connects sailors of all nationalities, even in time of war. Had Zschech investigated, he would have learned that the vessel that he just blew up is the *Ocean Justice,* and she is British. By some miracle, all fifty-six men on board are alive.

EIGHT

"I DEEPLY REGRET

TO INFORM YOU . . ."

light Sergeant Ron Sillcock is the ace pilot of Squadron #53 of the Royal Australian Air Force. In his 550 hours of flying experience, he has demonstrated a flair for feats of daring and bombing accuracy. He has two submarine notches on his airplane: *U-155* and *U-173*.

On November 10, 1942, Sillcock takes off from the US Army Air Force base in Trinidad on seek-and-destroy submarine patrol. The air base consists of two runways, known as Edinburgh and Xeres. A third landing strip, Tobago, is set aside for emergency landings. On this day, Sillcock departs from runway Edinburgh. Flying with Sillcock in his Lockheed Hudson III V925 light bomber twin-engine plane is an international crew made up of an American, a New Zealander, and three Aussies—all "good blokes," in Sillcock's estimation.

Thirty-year-old Sillcock is movie-star handsome, with curly blond hair and a firm jaw, hailing from a dairy farm in the southeastern Australian town of Gippsland. At the outbreak of the war, he enlisted in the RAAF because a knee injury made it painful for him to march in a land army.

The RAAF taught Sillcock mathematics and Morse code. He was selected for pilot training and sent to flying school in Southern Rhodesia. He has flown anti-submarine patrols off the coasts of Miami, Charleston,

Norfolk, Jacksonville, and Cuba. Now he is in Trinidad for an air patrol that is intended to last six hours. The weather is perfect.

Around 3:15 p.m., Sillcock is scanning the ocean for any sign of an enemy sub. The weather has turned cloudy, ideal conditions for hunting submarines because the thick gray blanket of clouds affords excellent air cover.

Then Sillcock sees her. It's a U-boat and she's about to become dead meat. He shuts down his engines, feathers the propellers, and glides in lethal silence toward his target.

On the deck of *U-505*, gunnery officer Gottfried Stolzenburg is scanning the skies for enemy aircraft. Through his binoculars, Stolzenburg is looking right at the plane. Does he mistake it for a shimmering bird in the tropical sky? Stolzenburg is about to find out that this bird has twin engines. Now the gunnery officer seems to recognize the menace coming his way. A shrill alarm is sounded. Emergency dive!

Too late.

Sillcock flicks on the engines. His Lockheed Hudson rumbles to life and swoops down full throttle. He lets loose with a 250-pound bomb. The aim is perfect, right on the foredeck.

Utter chaos ensues. The lights go out and every crew member is knocked off his feet by the blasts. To Hans Goebeler, it seems as if a giant fist has slammed into the boat. In the engine room, a jet spray of seawater pours in—the hull has been breached! The air starts filling with noxious smoke.

Zschech and his second in command, Thilo Bode, make their way up to the bridge to inspect the damage. They see Lieutenant Stolzenburg sprawled unconscious on the deck. Stolzenburg has been hit with shrapnel in the head and back, his body smeared with a mix of blood and seawater. The deck is a twisted mess of steel and metal. The antiaircraft cannon has been blown overboard and a thick strip of black oil is seeping into the ocean. The fuel tanks have split open!

This doesn't make sense. What the hell happened?

Off the starboard bow, through the smoke and fire, Bode can see the wreckage of Ron Sillcock's plane and a strange object floating in the water. Could that be a man? Bode blinks. It's not a body—it's a decapitated head. A blond head, hair flapping in the waves. For just a moment, Bode wonders if it could it be a member of *U-505*. Then Bode returns to the fuselage of the enemy plane, cut to pieces, sinking into the ocean, and he puts two and two together. The Allied pilot must have come in too low and dropped the bomb load before he was able to pull out of the dive. The Lockheed Hudson was brought down by the aircraft's own explosives. He looks at the blond head again. Is it the pilot? What a miserable way to go.

Peter Zschech goes into full panic mode. From topside, he sticks his head into the hatch and shouts the order to abandon ship.

The men don't know what to do. Do they obey the command, or do they remain at their posts? Such dithering reveals a scorn for a skipper that is inconceivable in the Kriegsmarine. From the control room, Chief Petty Officer Otto Fricke screams back at Zschech, "You can do what you want, but the technical crew is staying on board to keep her afloat."

Such defiance! Zschech flushes with indignation and deep shame. Has any U-boat commander ever been spoken to with such disdain in combat? But for once, Zschech has sense enough to defer to a more experienced officer.

"All right then, do what you can," he mutters.

Fricke and the engineering crew get to work. They plug the gaping hole in the hull with a rubber sheet and shore it with lumber. The seawater flooding into the engine room is sucked out by water pumps. Fricke switches air supply and the noxious fumes are blown out of the sub and replenished with fresh air. Topside, the deck crew begins the backbreaking task of clearing off the debris.

The next two days are filled with anxiety. The severely crippled *U-505* is utterly vulnerable to another air attack. Only one diesel engine is working, and she has no functioning antiaircraft weaponry. Nor can she dive. All the men can do is keep their fingers crossed that no Allied plane spots them.

One good stroke of luck is that the shattered top deck offers a good supply of scrap metal. Using an acetylene torch and welding machines and banging away with sledgehammers, the crew proceeds to patch up the holes in the hull. Finally, on the fourth day, Zschech is ready to conduct a test dive.

Fluten!—"Flood!" he barks. The crew can hear the water tanks gurgling with the rush of fifty liters of water pouring in. Slow and steady. Don't chance it. Now another fifty liters. And another. As the weight of water swells the tanks, *U-505* sinks lower and lower.

"Control room to all compartments: Watch closely for any leaks," Zschech howls over the intercom.

They go down twenty meters.

Then thirty meters—periscope depth.

Crack. Groan. The jury-rigged steel plates are holding. The men hang on. In just seconds they will know if they live or die.

Forty-five meters.

The pressure hole is buckling in! It feels as if they can't stop sinking.

That does it. Zschech orders the boat to rise and level out at thirty meters. For now, that's as deep as she's going to get. It means that the top of the conning tower is only a few meters submerged and under the right ocean conditions could be visible to any enemy plane flying overhead. For a submarine to go unseen from the air, she must be submerged at least thirty-five meters. But it could be worse. The crew is relieved. Cheers echo through the boat. It means they have a chance. At least they can dive to some degree.

That night, the radio operator erects a temporary antenna, and for the first time since Flight Sergeant Sillcock dropped his bomb on *U-505*, Zschech can send a message informing Admiral Doenitz's headquarters about the jam *U-505* finds herself in, and the urgent need for medical attention for Stolzenburg and the two other critically injured watch officers. Stolzenburg is hallucinating and coughing up blood. His breathing is labored. He has internal bleeding and probably punctured a lung.

A response comes back from headquarters several hours later. *U-505* is to rendezvous with a milk cow, *U-462*, off the Cape Verde Islands in fifteen days. Headquarters also offers some not particularly helpful medical advice: Keep the wounded men cool and give them good food. Other than that, they'll have to wait until the rendezvous.

U-505 proceeds easterly. During daylight, they stay underwater. Nothing can be done about the leaking oil slick she leaves in her wake. If any plane catches sight of it, all the pilot needs to do is follow the trail that will lead directly to the submarine. At night, *U-505* surfaces to recharge her batteries and quicken the pace across the Atlantic with her diesel engines.

On November 22, they catch sight of the potbellied milk cow *U-462*. Rings of salami and knockwurst and other food supplies are brought on board *U-505*. The fuel tanks are also topped off. Stolzenburg is transferred for lifesaving surgery to be performed by the medical officer on *U-462*.

On December 12, *U-505* limps into Lorient, having made the harrowing journey across Suicide Stretch, as the Bay of Biscay is now known. Greeting them is the usual marching band as well as hundreds of sailors, soldiers, nurses, and shipyard workers.

Everyone on the pier marvels at the battering sustained by *U-505*.

"Glad you made it home!" says Flotillenchef Viktor Schütze. "Hail to the crew of *U-505*!"

The U-boat fleet's chief engineer comes on board, inspects the hull, and can't believe what he sees.

"This is the most damaged boat ever to come back under its own power," he remarks.

The crew gather their seabags and report to their barracks. Superstitious as ever, they are left to wonder . . . if only their contemptible skipper Peter Zschech had thrown those flowers overboard ten weeks ago when they departed Lorient, this ill-fated patrol might never have happened.

Back in Gippsland, Australia, the father of Flight Sergeant Ronald Sillcock opens a letter from the Casualty Section of the Department of Air. He is informed that his son is missing in action:

I deeply regret to inform you that no trace was found of your son or other members of his crew. It was reported from one of the searching aircraft that a patch of oil had been sighted on the sea, but there was no sign of any debris.

NINE

CALL TO DUTY

I n Chester, Pennsylvania, everyone knows the Trosino family.

Mama Trosino is the owner of Trosino's Restaurant on West Fifth Street. It's the best Italian place in town. Mrs. Trosino is renowned for her old-country cooking, a menu that is basic but savory—eleven varieties of spaghetti, ravioli, veal scaloppini, chicken cacciatore. Also, some American dishes—but why would you want those? Most nights, the restaurant is packed to capacity. The kitchen is spotless.

Mama Trosino (friends call her Rose) and her husband, Angelo, an immigrant from a village northeast of Naples, have a big family, nine children. Angelo drives a produce truck, works in a quarry, and runs a small farm. Mama makes extra money sewing when she's not cooking and managing the restaurant.

Chester, Pennsylvania, is situated on the western bank of the Delaware River. One thing Mama always expressed was a desire that none of her five sons follow the temptation of the sea. So, of course, four out of five Trosinos became seafaring men.

Earl Trosino, the second oldest child, was enthralled by sailing vessels. Growing up in Chester, he watched foreign vessels dock at the port and unload product. "The water always drew me from the time I was a kid," he will recall many years later. "Going to sea is a man-made life, for you are going around with life in the palm of your hand. No one can realize the dangers out there until you see nature in all its fury."

While falling in love with the sea, an eight-year-old Earl Trosino also fell in love with his neighbor, Lucy DiRenzo. He can pinpoint the moment it happened: Lucy was playing in a tub of water in the backyard with her sister. She was five, and when Earl saw her, that was it. Lucy was eight when Earl proposed. True story. All through Chester High School, they held hands.

Earl became a cadet at the Pennsylvania Nautical School, learning basic seamanship on the USS *Annapolis*, an old Navy gunboat that serves as the school ship. One day in 1927, Lucy played hooky from school. On that same day, Earl jumped ship. By prearrangement he and Lucy met at the train station and eloped to Maryland with forty dollars in their pockets, borrowed from friends. A taxi driver took them to the local parson and earned a few bucks bearing witness to the marriage. They kept it a secret— one that only lasted a year after Lucy's mother opened her mail and saw a letter written by young Earl Trosino addressed to, "My dear Wife . . ."

Right then, Lucy's mother told her to march down to her father in the basement, kneel, kiss his hand, and beg forgiveness. Which Lucy promptly did.

It's not that they objected to Earl at all. They just wanted a big church wedding for their Lucy, which was finally accomplished on September 23, 1928, at St. Anthony's in Chester.

When the bottom dropped out of the shipping business during the Great Depression, Trosino found himself "beached, hungry and with a wife to support." He made money from a local contractor—twenty-five cents an hour shoveling dirt and cleaning bricks. Then the Sun Oil Company offered him steady employment. Over the next decade, he sailed on Sun Oil tankers. During all that time, Lucy accompanied him to the dock for the start of another long voyage at sea, but never waited for his ship to depart.

"I always knew he would come back," she explained.

Trosino was also a merchant marine reservist, and he had no illusions about where he stood in the US Navy Reserve hierarchy—"at the ass-end." But in his opinion, "we're the only real seafaring men in the whole damned outfit."

As the winds of war loom, Trosino is called to active duty as a full lieutenant in June 1941—six months before Pearl Harbor—and assigned to the attack cargo ship USS *Alcyone*, then the aircraft carrier USS *Long Island* as an engineering officer. He participates in historic events: the Battle of Midway and the Guadalcanal campaign. An explosion in the engine room on the *Long Island* sends him to the naval hospital in San Diego where, he likes to joke, he went in for a "general overhaul" because the Navy doctors end up operating on his hernia, tonsils, and varicose veins. He cracks wise about the experience:

"When they finished using the body of one Earl Trosino as a pincushion, whetstone and sounding gauge, a medical survey was held on the remains."

But the Navy isn't through with him yet. Trosino is still recuperating when he's ordered back to active duty. Wobbly legs and all, he is discharged from the hospital—and he understands why. They need all available beds for the casualties coming in from the Pacific theater of war.

The Navy dispatches him to a new ship, the USS *Guadalcanal*, where he reports to Captain Daniel Gallery as chief engineering officer.

A full-blooded Italian American, taking orders from an Irishman. This should be interesting.

TEN

SABOTAGE

The directive from the British War Cabinet lands with the subtlety of a depth charge. It orders Air Marshal Arthur Harris—"Bomber Harris," as the commander in chief of RAF Bomber Command is known—to target all U-boat bases in Germany and occupied Europe and attack them to "maximum scale."

First on the priority list of cities to bomb out of existence is the French seaport of Lorient.

Leaflets are dropped over Lorient warning the civilian population of the impending blitz and notifying them there is still time to evacuate.

Anyone who ignores the warning finds out the hard way that it wasn't a bluff on the night of January 14, 1943, as ninety-nine British planes drop seventy-six tons of bombs and incendiary containers filled with magnesium and thermite on Lorient. It's a living hell as more than eighty fires break out—and that's just for starters. The next evening, the Brits come back with even greater forces—140 tons of bombs, triggering four hundred fires. American planes take over on the third night with a sortie of B-17 bombers that destroy forty buildings.

Hans Goebeler and his *U-505* comrades are shaken when they see the devastation downtown, and they're jolted again when they turn the corner and step into the red-light district. It's intact! The bars and brothels are still standing. The streets sparkle with beautiful mademoiselles tempting the sailors and the sweet scent of perfume mixed with rouge and lipstick, even more intoxicating than before because it's totally unexpected.

A delighted Goebeler steps into his favorite whorehouse. The madame remembers him. She helps him off with his bulky leather coat, and Goebeler tosses his fur cap into the air, landing it right on a hook. He's bursting with confidence as he asks about Jeanette. She hasn't been seen in weeks, the madame informs him. She left the city without telling anyone and nobody knows where she is or what happened to her. Too bad. In no time, Goebeler has a drink in his hand and a French prostitute on his lap. The war has hardened his heart. Screw Jeanette. He downs the drink. And another.

Dawn rises. Goebeler leaves the brothel and starts making his way back to the barracks. It's been a blissful night, but the sudden wail of an air raid siren screeching across the sky shocks him back to reality. British bombers are about to deliver another message to the Nazi occupiers.

Antiaircraft batteries commence firing and, as they do, Goebeler stumbles upon an underground shelter and sprints down twenty steps, but a thick iron door is slammed in his face. There's no more room. Goebeler curls into a fetal position in a corner and waits it out, as bombs rain down above him and the shelter walls shake violently.

■ ■ ■

The battle-battered *U-505* is lifted into dry dock for extensive modifications and repairs. The entire crew is issued three weeks leave, and they all go their separate ways.

Radio operator Karl Springer heads to his village in Germany near the Polish border to visit his mother. Springer's father and seventeen-year-old brother, Herbert, are fighting on the Eastern Front. But Karl also has a second mission in mind. He crosses the border and ventures eleven miles into Poland, to a garden restaurant in a garrison town that caters to six thousand German soldiers. His girlfriend, Isle, works as a waitress here, and she's thrilled when Springer shows up—doubly so when he gets down on one knee to propose. She says "yes," and the two are officially engaged.

Torpedoman Wolfgang Schiller really lives it up on his leave. He ends up in Bad Wiessee, a spa town on the north side of the Alps in Upper Bavaria, famous for its sulfur baths. Bad Wiessee is the official municipal "sponsor" of *U-505*, and Schiller is welcome to stay at a swanky hotel and enjoy recreational boating on Lake Tegernsee, where Hitler himself owns a house.

Hans Goebeler boards a train for his hometown of Bottendorf for the first time in two years. He is "proud as a peacock," a golden submarine insignia affixed to his breast pocket. *Just wait 'til they see me . . .*

The journey by rail takes him first to occupied Paris, then on to Frankfurt. Along the way Goebeler encounters the enchanting villages that dot the German landscape. This is fairy-tale Germany, with picturesque, timber-framed farmhouses, medieval church spires, and Old World charm. But it breaks his heart to see the drubbing the British and Americans are delivering to the Fatherland. Even with the fall of Stalingrad and the defeat of Rommel's Africa Korps, Goebeler's faith in ultimate German victory remains unshaken.

The final stop is the train depot in the Hessian town of Frankenburg, which is a five-minute drive from Bottendorf. Like the rest of Germany, Frankenburg has fully embraced Hitler and fascism. Stone Lane (*Steingasse*) has been renamed Adolf Hitler Street by the town council. Roddenauer Street is now Hermann Göring Street, after the obese Nazi Luftwaffe commander. The town's Jewish population is long gone, having fled due to systematic persecution or having been deported to death camps.

Seabag slung over his shoulder, Goebeler walks into his house like a conquering hero and is greeted with a joyous welcome from his parents, Heinrich and Elizabeth, and his sisters, Anna Marie and Kati. The boy they knew is now a man.

Heinrich Goebeler knows what his son is going through, having personally experienced the horrors of war. During World War I, he enlisted in the German Army at age eighteen and fought on the Eastern Front before he was captured in 1914 and sent to a slave labor camp in Siberia. Heinrich didn't return home until 1921, seven years after he was taken prisoner. Heinrich Goebeler despised the chaos of the Weimar Republic, and it is his fervent belief, which he has drummed into his son, that Adolf Hitler and National Socialism have restored order and greatness to Germany.

All the neighbors are delighted to see young Hans again. What a credit to the Goebelers! In appreciation of Hans's military service, they come bearing gifts: sausages and smoked meats and baked goods in large quantities, to bring back to the boys on *U-505*. Generous gifts indeed, considering food rationing.

When the younger Goebeler's leave is over and it is time to go, his mother bursts into tears. Will she ever see her Hans again? Heinrich Goebeler maintains a steely composure. Hans must bid farewell. Duty calls, but the long goodbye puts his journey in jeopardy—he's going to miss his train! Luckily, his father works for the Reischbahn, the German National Railway. He uses his connections to call in a big favor, pleading for a delay in its departure, and is granted three minutes. Such is the best the railroad can offer in a nation that prides itself on making the trains run on time, but it's enough for young Goebeler to arrive at the station and jump on board just as the train is pulling out.

When he returns to Lorient, Goebeler is taken aback. His cherished boat *U-505* is almost unrecognizable under a fresh coat of dark gray paint, and that's just the outside. Inside, the boat has been upgraded with the latest German technology and armaments since she limped into port six months ago. She is outfitted with a larger conning tower and state-of-the-art antiaircraft weaponry on the deck. She has also been demagnetized to reduce the risk of hitting a sea mine. She is ready for battle as never before.

The pier is bustling with activity, with truckloads of torpedoes and provisions being brought on board.

Goebeler figures he is due one last hell-raising night on the town. The relentless Allied bombardment has flattened most of the city and the population has fled. There were forty-six thousand French citizens before the war, falling to just five hundred people now, stumbling around dazed amid the ruins. But the red-light district endures. Goebeler is scrutinizing the "tasty dishes" of Lorient when he sees an attractive young woman heading in his direction. There is something familiar about the way her hips sway with each step. Could it be? . . .

Yes! It's Jeanette. She throws her arms around him and gives him a warm smile when she notices he's still wearing the St. Christopher's pendant she once gave him for safe travels. A flood of questions come pouring out. Jeanette explains her disappearance this way: It seems she was driven out of Lorient by the Maquis, as the French Resistance is known in this region of the country. Her fraternization with the enemy has made her a target, and she had to flee for her own safety, but she is back, for now.

They spend the night together. The next morning, Jeanette gives Goebeler a few packs of cigarettes as a parting gift. Then she informs Goebeler she must leave and start a new life somewhere east, where they don't know about her history.

Goebeler returns to his barracks, sad to part ways with Jeanette. Who knows if they will ever see each other again? At the farewell banquet that night, there are many rounds of toasts. Everyone gets drunk. The men sleep it off until ten thirty the next morning, at which time a bus transports them to the pen where *U-505* is docked. Peter Zschech awaits them. Even the martinet captain is suffering a hangover. Goebeler conceals two dozen lemons inside his personal belongings. Those lemons will stave off scurvy, and he's not the only *U-505* sailor with bulging pockets filled with citrus to be smuggled on board.

About fifty well-wishers and dignitaries from the Second U-boat Flotilla are there to see them off. But this time there is no pomp and circumstance, no booming military band playing patriotic German marches. In its place, just a troupe of accordion players. The crewmen—with newfound, firsthand knowledge—grasp that they are embarking on a mission from which, like so many of their comrades, they may never return.

On July 3, at eleven at night, in a drizzling fog, Kapitänleutnant Zschech orders the lines cast off. Zschech seems unsettled. One reason may be the absence of his closest comrade, Thilo Bode, who is no longer with him. Bode has been promoted to command his own boat, *U-858*.

Zschech's new second in command, or first watch officer, is twenty-six-year-old Paul Meyer. Unlike Zschech, whose father was a Navy doctor, Meyer has risen through the ranks the hard way, without any naval family pedigree. Nor did Meyer graduate from the navy academy. He enlisted in 1936, served on a destroyer, where his talents were recognized, and transferred to the elite U-boat service. The men on *U-505* find him a welcome relief from the tyrannical Bode, who seemed to have reinforced Zschech's worst tendencies. Meyer is more like one of them.

U-505 clears the pen and cruises at a half knot speed to the mouth of Lorient Harbor. There, they join a flotilla of seven minesweepers and four U-boats that are also departing on their military missions. Taken together, the U-boats and minesweepers form an effective antiaircraft defense canopy as they traverse Suicide Stretch. Clearly the Germans have copied the convoy tactics from the Allies. The Nazis have learned there is strength in numbers.

But there's trouble right away.

Barely an hour into the voyage. *U-505*'s diesel engines sputter. They shut down entirely around midnight. One minesweeper stays behind to provide air cover while the other U-boats and their escorts continue onward. It takes ninety minutes of repair work for *U-505*'s diesels to kick back into gear.

As *U-505* thunders on a full power run out of the Bay of Biscay, it's time for a test dive. All goes well, until forty meters down, when a leak springs in the starboard propeller shaft. It's as deep as Zschech dares descend.

Five days out of Lorient, submerged at forty meters near the Spanish coast of Cape Finisterre, Zschech orders periscope depth. Through the eyepiece, he scans the sea and sees twilight has come.

Christ. Three British destroyers, dead ahead. Coming this way.

"Quick!" he cries. "Take us down to sixty meters."

Six depth charges are dropped right on top of her.

"One hundred twenty meters!" Zschech orders. "Emergency lights!"

How could the enemy be so on the mark? It takes thirty-six hours for *U-505* to shake the Brits.

It's one glitch after another. Bad luck? Coincidence? Zschech smells a rat. He suspects sabotage. The dockworkers back at Lorient are made up of ethnic Germans recruited from Poland and other Eastern European nations that are now occupied by the Nazis. Their loyalty is open to question. After all, they are a conquered people. There are also many Frenchmen among the dockworkers. It could be the handiwork of the French Resistance. There shall be hell to pay for this, Zschech vows. At this point, repairs at sea are impossible. Zschech has no choice but to return to Lorient.

Once again, Zschech brings *U-505* up to periscope depth and peers into the periscope.

"Holy shit!"

An enormous oil slick of diesel fuel is leaking out of *U-505*, creating a giant glowing rainbow that pinpoints *U-505*'s precise position. Damn those saboteurs.

It's a miracle that *U-505* pulls safely into the submarine pen back at Lorient, especially after the flotilla engineer and shipyard officials come on board to inspect all the seals, diving tanks, fuel tanks, and battery cells. Their conclusion? Somebody has poured battery acid on the seals. Other U-boats are experiencing similar acts of sabotage. One boat is found to have sugar in the lube oil, another a dead dog contaminating the drinking water tank.

Repairs commence on August 1, 1943, under tight security, as the Nazis keep a watchful eye on the shipyard workers. They also retaliate against the French dockworkers. Half a dozen men are put to death by firing squad. Were they guilty as charged? Or the victims of Nazi retaliation based on collective guilt? Nobody can really say.

Two weeks later, *U-505* sets out again. A test dive is conducted. But at thirty meters, the crew hears an ear-shattering metallic cracking. Is it the settling of the metal plates, or a serious structural issue? A deeper dive is conducted to nail down the problem. At sixty meters, they hear hissing and gurgling, then an alarming boom. It sounds as if water is penetrating the pressure hull.

Back to Lorient. An inspection reveals the pressure hull seams have obviously been tampered with. The steel plates are supposed to have been welded together with a blow torch, but these plates have been sealed with strips of substandard oakum, which is nothing more than tarred fiber—hardly the stuff to withstand the pressure of a deep dive. Two more weeks of repair are required.

U-505 casts off once again. A test dive of sixty meters depth is conducted just offshore. Suddenly, vibrations break out, and more gurgling. The main induction valve has buckled. Another case of sabotage! Will this never end? Zschech turns red with fury and orders the boat back to Lorient. It's getting to be a joke. They didn't get farther than three hundred meters out of the dock.

They try once again a week later. The test dive goes well. It's a promising start. Zschech dives to periscope depth and looks around. *Oh no.* Another diesel fuel leak! Then an engineer reports to Zschech that the lube oil is contaminated with sugar.

On returning to Lorient, another shocking discovery: A saboteur has drilled a pencil-sized hole into the fuel tank. How the hell can this keep happening?

The Nazis get to thinking. Perhaps it isn't the French Resistance after all. Gestapo agents come on board *U-505* and start investigating the crew. Could they be deliberately crippling the boat? Are there any enemies of the Third Reich among them? Radio operator Karl Springer is stunned to find himself under interrogation by the dreaded secret police, apparently for no reason other than he happens to be around.

"You're committing sabotage!" the Gestapo agent screams. "You don't want to go out anymore!"

"It's not true!" Springer shouts back.

The agent keeps at it. "Don't you want to be out there?"

Springer explains from the point of view of the crew. "We are good Germans, loyal to the Third Reich," he says. "But we're not going out there with a defective boat."

Even the thick-skulled Gestapo get Springer's point. What crew member would be stupid enough to sabotage a boat when that could get them killed? The Gestapo realize there's nothing to the accusations and drop the investigation.

Zschech and *U-505* give it another shot on September 18. So far, so good. Perhaps *U-505*'s luck has turned around. The skipper is cautious to the extreme as he creeps out of the Bay of Biscay, coming to the surface for about ten minutes in the morning for a quick reading of the sextant, and then remaining underwater except for three hours a day. Submerged, he proceeds at a snail's pace, hoping to avoid being spotted by Allied air patrols. But two days out, a cylinder in the port diesel engine freezes. The crew manages to fix it themselves. Three days later, the trim pump breaks down. *U-505* turns back. No one onshore is surprised to see the submarine pulling in.

October 18. *U-505* is cleared to depart. No military band. Not even the accordion players. This time it's a lone musician playing the harmonica. Could there be a more pathetic sound to ship out by?

Surprisingly, the test dive goes off without a hitch. Not a single issue.

Five days later, *U-505* clears the Bay of Biscay and enters the deep Atlantic. A collective sigh of relief, from stem to stern.

Zschech opens his instructions and reads them to the crew. *U-505* is ordered to the Caribbean. At the Azores, Zschech turns westward. After ten frustrating months, the crew is finally back in the thick of the war.

ELEVEN

USS *CAN DO*

aptain Dan Gallery is hardly bowled over when he sees the USS *Guadalcanal* for the first time. It's frail, homely. The *Guadalcanal* is 512 feet long, about half the size of the great ladies of the Navy— majestic aircraft carriers like the *Essex, Yorktown, Intrepid,* and *Hornet*— and considerably slower. She also carries fewer planes. She reminds Gallery of Cinderella—that is, *before* the fairy godmother waved her magic wand.

The *Guadalcanal* was built on the cheap at the Kaiser shipyard in Vancouver, Washington—one of fifty similar ships known as "Jeep carriers," or "escort carriers," or "baby flattops." The objective was to churn them out on a mass-production scale and a tight budget by adopting the assembly line principles of the Ford Motor Company.

The rumor circulating around the shipyard is that the *Guadalcanal* is so structurally unsound she'll break in half the moment she casts off. There is no faith in the thousands of farm boys, shoe clerks, soda jerks, high school girls, and all those Rosie the Riveters who welded and riveted the ship into a seaworthy machine of war.

But there is something about the ship that greatly appeals to Gallery's appreciation for history, something that no other ship has, and that is the name. The Navy had offered Gallery command of another ship that had an urgent need for a new skipper, but Gallery insisted on sticking with the *Guadalcanal.*

True, it may lack speed and thick steel plates and watertight compartments, but naming it after the epic battle that was turning the tide of war against the Japanese in the Pacific—that was the ship for Dan Gallery.

"*Guadalcanal* is my baby," he says. He wouldn't swap that ship for anything—well, maybe the *Saratoga.*

The crew of the USS *Guadalcanal* reports for duty. Most of the lads are fresh out of six weeks of Navy boot camp and know next to nothing about aircraft carriers or, for that matter, life at sea. They have a better understanding of cornfields than blue water.

Gallery keeps the rules simple. Stick to your post in an emergency, he tells his men, even when everything starts to stink like hell. If the ship gets hit, put out the fires. And remember that you can get out of most jams as long as the water doesn't get higher than your waist.

Each sailor is handed a memo with these words, written by their captain: "The motto of the *Guadalcanal* will be Can Do, meaning that we will take any tough job that is handed to us and run away with it. The tougher the job the better we'll like it."

There is also a postscript: "This ship will be employed on a dangerous duty. We will either sink the enemy or get sunk ourselves, depending on how well we learn our jobs now and do our jobs later. *Anyone who prefers safer duty see me and I will arrange to have him transferred.*"

After that memo, the crew starts calling the *Guadalcanal* by another name. Henceforth she is known as the USS *Can Do.*

The crew also has a nickname for Gallery—"Full Flaps," for his protruding ears.

A bottle of champagne goes flying into the ship's bow on June 5, 1943. She is officially underway.

When Gallery boards the *Guadalcanal*, he finds the captain's quarters "the most primitive I've ever seen on a ship that size." There's barely enough room for him and his dog, a feisty border collie mix named Flairby that Gallery adopted in Iceland who now accompanies him on his new vessel. Yep, Gallery brought his dog. Rank does have its privileges.

One morning, Gallery gets to chatting with the chief boatswain's mate, a man who has fifteen years of Navy service behind him. What do you think of this new class of seamen coming out of boot camp, Gallery asks.

The chief doesn't sugarcoat it. "Cap'n, I'd swap 'em all for a bucket of oil rags."

Maybe so. It's true, they are so damn green, but there's something about these American lads that Gallery finds compelling, and that is their uncanny capacity to figure out how to operate machinery and make things work. They seem to be natural-born grease monkeys, with a knack for gadgetry and equipment that's as useful at sea as it is on dry land.

■ ■ ■

October 15, 1943.

The time has come for *Guadalcanal*, now fitted out and loaded with provisions, to cast off on her shakedown cruise, work out the kinks, and see how she takes to salt water. It just so happens that the first night, as the ship noses her way down the Columbia River, she runs smack into the tail end of a gale. Inside of fifteen minutes, white-capped waves eighteen feet high are crashing onto the flight deck. With each heave and roll, the steel templates pop in and out. *Plop-Plop! Plop-Plop!* They sound to Gallery like thunderous "cracks of doom," almost like squeezing a gigantic can of oil. Considering those rumors back at the shipyard, the crew starts wondering if the *Guadalcanal* really is structurally defective. Could she split in half? But the "old gal," as Gallery takes to calling her, holds up fine in the storm.

The ship sails south, on her way to San Diego, her first stop on the shakedown cruise.

Gallery has several strong senior officers to help get *Guadalcanal* shipshape.

The executive officer is Commander Jessie Johnson. Gallery considers him a "tower of strength." The chief engineer is Earl Trosino, and he's another story. In those early days on the shakedown cruise, Gallery and Trosino butt heads. Maybe it's just the familiar push-pull between Italian Americans and Irish Americans. Trosino calls Gallery the "Fighting Irishman," and who knows what Gallery mutters under his breath about

Trosino? Gallery is a stickler for regulations, and Trosino, who comes from the more casual customs of the merchant marines, sometimes neglects to salute him. When Gallery reads Trosino the riot act, Trosino, who is just as hardheaded, doesn't appreciate the dressing-down and tells Gallery just how he feels. Not many officers speak to Gallery that way. Trosino actually asks for a transfer to another ship, but the paperwork keeps getting "lost" by the XO.

Quickly, however, Gallery comes to appreciate Trosino's extraordinary knowledge of ships.

There are lots of mechanical problems with the *Guadalcanal* to iron out. Everything seems to be breaking down.

"This is a hell ship," Trosino blurts out.

When the boilers go haywire and send smoke pouring over the deck, Trosino has had enough. He marches to the bridge.

"We have to shut down," he tells Gallery.

The captain has had enough of the whining. "Trosino, is this gonna run or isn't gonna run?" he explodes. "If I were another skipper in this man's Navy, I'd have your hide hanging on my bulkhead."

Another time there is a problem with the starboard condenser. Trosino starts fiddling and finds a pair of pink panties stuffed inside. He fishes them out and carries them to the skipper like a lab specimen, dangling them between his thumb and forefinger. How the panties got there, nobody knows. Maybe Rosie the Riveter.

Just as he did in Iceland, Gallery pays special attention to morale and the creature comforts on the *Guadalcanal*. Athletics are a big deal. Basketball tournaments on the hangar deck. Touch football. Boxing and wrestling matches in a professional ring, with ropes. A gym with parallel bars. Catching rays when the sun is out. Or just playing catch.

Gallery makes a point of dropping in unannounced to snoop around the galley and bakery shop. The way he sees it, the Navy committed to providing his boys three square meals a day, and he's going to see that they get them.

The ship's library is loaded with four hundred records and plenty of checkers and decks of cards. And every evening at 6:00 p.m., his XO, Johnson, who has a way with words, hosts a radio show played over the ship's PA system, reporting the news of the day just like the famous broadcaster Lowell Thomas. Johnson even delivers mock commercials plugging the *Guadalcanal*'s laundry service and barbershop. When the news is over, he spins a few "hot platters" on the record player.

Once a week or so, Gallery steps in front of his eight hundred sailors so they get to know him and he them. As they prepare to enter combat, the message he wants to deliver is that it's natural to feel scared. He gets scared, too, notwithstanding those four gold stripes on his shoulders.

Another routine Gallery brings to his ship is a daily devotion. Right after sunrise, the boatswain's mate blows his whistle to announce morning prayers. All hands stop what they are doing and doff their caps, face the bridge, and pay attention as the ship's chaplain, Father Christopher Weldon, delivers a thirty-second nondenominational prayer over the loudspeaker.

The *Guadalcanal*'s chief medical officer is Dr. Henry Morat. He speaks English (and several other languages) with a thick French accent, and Gallery finds him to be quite the flamboyant character. There is also a dentist on board.

En route to San Diego, *Guadalcanal*'s first medical emergency is declared. A sailor is doubled over in agony. Dr. Morat diagnoses appendicitis. The good doctor prepares to operate, and Gallery does his part by adjusting speed and reducing the pitch of the ship.

About a half hour later, Dr. Morat, still wearing a surgical gown, comes to the bridge carrying a bloodied bundle of gauze bandages. He shows Gallery what's inside—a gangrenous appendix.

"I got it just in time," he says with pride.

First, Chief Engineer Trosino shows Gallery the pink panties, now Doc holds up a human organ. *Hell's bells*, Gallery thinks. *Do these guys have to show me everything?*

San Diego at last! The shakedown cruise is over. Turns out, despite all the mechanical issues that are driving Trosino nuts, Rosie the Riveter can build a pretty solid ship.

On Halloween, the air group reports for duty—thirty-one officers and forty-seven enlisted personnel. Five FM-1 Wildcats and five Grumman TBF-1 Avengers are hoisted on board. More planes to come.

The squadron commander is Lieutenant Commander Joseph Yavorksy, a thirty-year-old Naval Academy graduate from Youngstown, Ohio. Another pilot is Lieutenant Phil Berg, and many must wonder what the hell he's doing here. For one thing, he's no kid, answering his nation's call to arms at age forty-one. For another, he's not off the farm—he's a millionaire talent agent who created the concept of the "package deal" and represented Clark Gable, Judy Garland, Frank Capra, and other Hollywood big shots.

Another flyboy with the lowly rank of lieutenant junior grade has the moxie to bring along his cocker spaniel, a cute little thing named Prissy. She can beg, play possum, and roll over on the command "snap-roll."

Confidence: That's what makes these glamor boys of the Navy a breed apart from the regular sailors. One sore point is that they always seem to score the prettiest ladies at the USO parties. Plus, the pilots earn 50 percent extra hazard pay, which they deserve because landing on a flat-top is probably the riskiest duty in the Navy. And they're exempt from routine watches. Friction with the grunts is inevitable, one reason why Gallery is determined to make the *Guadalcanal*'s first takeoff and landing himself. Not that he's a hot dog, he only wants to prove the point to the other pilots that he's not just captain of the ship but also a capable aviator like them.

He climbs into an advanced trainer—a North American SNJ-4 Texan—revs up the engineer, braces himself, and in the next moment is catapulted off the flight deck with a jolt. The flight is brief—a single swing around the ship—then he lands to the cheers of his men on deck. Point made.

On the third day of qualifications for aircraft carrier landings, there is a mishap. A Wildcat piloted by Lieutenant (junior grade) Thomas Dunnam Jr. hits the deck, misses all the landing wires, and bounces until it comes to a dead halt at the barricades. Not a textbook landing.

Then an Avenger piloted by Ensign Richard Law is returning from routine anti-sub patrol when the plane's arresting hook fails to lower. All Ensign Law's Navy training comes into play. He turns off the battery switch. He hits the circuit breaker. Nothing works. He pulls on the emergency release. It's stuck. Time to brace for an emergency landing. First, he jettisons his bombs and tries everything one more time to get that arresting hook to deploy. He throws the battery switch on and off. He violently shakes the plane. He flicks the hook switch up and down. He pulls the emergency release one more time. Nothing. He's out of options. After an hour of effort, he's given the okay by the *Guadalcanal* to make a hard landing.

His first approach—he gets a wave-off.

His second approach—he hits the deck. There's a bounce, and he crashes into all three barriers, nose up. Immediately he turns off the fuel, and he and his two crew members hustle out of the plane, rattled but alive.

The flyboys are a wily bunch. Lieutenant Berg arranges for a bus to take Gallery, the ship's doctor, the dentist, and all the squadron officers on an excursion to Hollywood. First, they make a pit stop in Beverly Hills and check out Berg's mansion on Sunset Boulevard. Nice digs. Not bad for a lieutenant jg. They also meet Berg's wife, the beautiful film actress Leila Hyams, the heroine from *Island of Lost Souls* who turned down the role of Jane in the movie *Tarzan the Ape Man*. They stay several nights at the Beverly Hills Athletic Club. Thanks to the well-connected Berg, an entire section of the Florentine Gardens nightclub is reserved for them. They dance the night away in a conga line with pretty starlets. The next day they lunch at Paramount Pictures studios where they watch a scene being filmed with Dorothy Lamour, who is making the musical comedy *Rainbow Island*. Funnily enough, it's about three torpedoed merchant marines who find themselves stranded on an isolated island in the South Pacific inhabited by beautiful native women. Hey—it could happen!

Berg saves the best for last.

That night, at Marcelle La Maize's, there is an open bar. And what a crowd! There they are, making small talk with Olivia de Havilland, Cecil B. DeMille, Harpo Marx, Loretta Young, and other important Hollywood names. Captain Gallery dances until midnight. The gossip columnist Louella Parsons happens to be there and sets her poison pen aside, writing it up as "one of the nicest parties ever given in Hollywood." Gallery can only hope his wife, Vee, and the Navy brass back at the Pentagon don't read about it!

These are good times for Gallery. He's finally off his Icelandic shore assignment and back on blue water, in command of an aircraft carrier during the most epic war in world history. Naturally, now is the perfect time for the Navy to drop the other shoe.

When the order comes through, Gallery's heart sinks. The *Guadalcanal* is assigned to the Atlantic theater, which means the cornerstone of its mission will be escorting merchant ships and undertaking tedious anti-submarine patrols. What a grind. No opportunity for great sea battles like Midway. No island invasions like Guadalcanal. It's a huge letdown for Gallery. After Iceland, he thought he was done with the Atlantic. The Pacific is "where the good Lord means me to be."

The day after Thanksgiving 1943, the *Guadalcanal* passes through the Panama Canal. And when she is on the other side, Gallery is back in the Atlantic.

TWELVE

OVER THE SIDE

Kapitänleutnant Peter Zschech puts his Luger service revolver to his temple with a trembling hand and braces himself. Just a bit more pressure on the trigger, and it'll all be over . . .

Chaos all around. The lights are out, and a barrage of depth charges has his submarine flailing. Will Zschech end his life, or will he drop the gun? And how has he reached this desperate crossroads?

The nightmare he's living begins at around 11:00 a.m. on Sunday, October 24, 1943. *U-505*, submerged, is plodding along to the steady hum of electric motors, making sluggish progress across the Atlantic on her way to the Caribbean Sea for war patrol. The men are shivering in their bunks as droplets of cold condensation drip onto their faces and soak their clothes and blankets. The boat reeks with the smell of rotting potatoes mixed with the rancid body odor of un-showered men, and the stench of the shit bucket in the engine room. A "cold clammy coffin" is Hans Goebeler's assessment.

Zschech will have to resurface soon, recharge the batteries, and air out *U-505*. But right now, it's just too dangerous. Even underwater, they can hear the distant boom of depth charges exploding. It can only mean that another U-boat up ahead is getting pounded by the forces of the British Navy. It continues like this for six harrowing hours. At a loss for what to do, Zschech withdraws into his cabin and pulls the curtain shut. Is he pondering his next step? Or is he shutting himself off because he has no idea how to get out of the jam he's facing? Has he lost all nerve? The crew is mystified. What are they supposed to do?

At sunset, the depth charge explosions become more thunderous. They're getting closer. Suddenly, the radioman hears the steady drone of an enemy ship's propellers on the surface. Must be a British destroyer. Zschech has to be informed, even if it means being on the receiving end of an irrational tirade. The radioman heads to Zschech's cabin and gives the skipper the bad news.

It's like talking to a ghost. All the color has drained from Zschech's face. He finally emerges from his cabin and does something so bizarre nobody can figure out what it means: He starts climbing up the ladder into the conning tower. Watching all this in the control room, Hans Goebeler exchanges looks with his crewmates. What's Zschech doing up there? Has he gone mad? The only rational reason for being there is to use the periscope, but *U-505* is cruising at a depth of one hundred meters. At the current depth, the periscope is useless.

All at once . . .

Ping. Ping. Ping.

The sound is so sharp it can be heard by everyone on *U-505*. Their hearts drop when they realize their boat is getting scanned by ASDIC, the new sound technology developed by Allied scientists. ASDIC is named after the Anti-Submarine Detective Investigation Committee. It emits a signal at regular intervals, and when the sound hits a solid metallic body like *U-505*, it bounces back as an echo. Later, the world will come to call this breakthrough "sonar."

Each *ping* is getting shorter and sharper. It means the ship above them is zeroing in on *U-505*'s precise position. The enemy is now directly above the submarine. And still nothing from Zschech. Everyone is waiting for orders from the skipper. There must be some evasive maneuver they can take. At least dive deeper. Slow speed. Cease all operational activity. Enter silent mode. Send all nonessential personnel to their bunks. Keep the enemy guessing. For Chrissakes, do *something*.

Too late.

Boom!

A depth charge—right on top of them.

The men are thrown off their feet.

The boat shakes wildly.

Charts, instruments, and shards of glass fly across the control room.

But depth charges are never dropped just once. There is always a sequence.

Here they come.

Boom!

The lights go out. The pressure hull rings "like a church bell."

A frantic Zschech climbs down the conning tower ladder.

Boom!

Boom!

More depth charges. Then the mightiest blast of all. It almost rolls the U-boat over. Sailors fall headlong into one another.

Zschech hangs on to the bulkhead and recovers his balance. Perhaps, in these final seconds of his life, he's thinking of his comrade in arms Thilo Bode and how things might have turned out if only he had Bode to cling to in this crisis.

Zschech removes the sidearm pistol from his leather holster.

His semiautomatic Luger is a high-grade product of German engineering. It is beautifully fitted, with a finely checkered walnut grip, designed to be fired with one hand. It carries eight rounds and weighs just one pound, fifteen ounces. There is a swastika and Nazi war eagle stamped on the highly polished metal.

Zschech clicks the safety. No one seems to be paying him any attention. He points the 4.7-inch barrel to his temple. Then he pulls the trigger.

A crisp retort rings out.

At first, the crew doesn't think much about it. They're holding their breaths because the depth charges have stopped for the time being. Not necessarily good news, because it must mean the British destroyer—if it *is* British—is reloading her depth charges and circling around for another go at it.

Suddenly the emergency lights come back on, revealing a sight that stuns Hans Goebeler. It's the skipper, face down, a pool of blood trickling out of his brain, spreading down to his legs.

The soundman struggles to turn Zschech over. Then other men grab his limbs and carry Zschech to his cabin. Goebeler creeps up to the cabin and peers inside. It's Zschech all right. The skipper is sprawled on his bunk with a small bullet hole in his temple. With each beat of his heart, a gush of blood flows from the wound. More men gather to watch. They listen to a death rattle gurgle. Hard to believe, but Zschech is still breathing.

The death rattle is getting louder. The crewmen look at one another. The destroyer on the surface is listening for any sound. Did the skipper doom them all with that booming gunshot? Absolute silence is necessary. Then somebody takes a pillow and smothers Zschech. Four powerful hands join in and press hard on the pillow. Are they trying to muffle the death rattle? Or is it a mercy killing? Are they hastening the inevitable? Maybe all three.

The U-boat's doctor, Friedrich-Wilhelm Rosenmeyer, tries to pull the pillow away. They are murdering the skipper!

Paul Meyer, the executive officer, tells Dr. Rosenmeyer to quiet down.

"There is nothing you can do for him now," Meyer hisses. "Those ships up there are trying to send us to hell. So, please, doctor, be quiet."

Rumors spread throughout the boat like a drumbeat, passed from sailor to sailor until it reaches the far end of the boat, the torpedo room. *Zschech shot himself!*

Radio operator Karl Springer runs into the chief machinist and is shocked to see that his hands are stained with blood.

"Chief, did the skipper shoot himself?" Springer asks.

"Where the hell did you get that?"

Springer plays dumb. He doesn't want to be accused of rumormongering. "I dunno."

But the chief machinist realizes something like this can't be covered up. Everyone will know soon enough. Plus, there's the dead giveaway—his blood-soaked hands.

"Yes," he admits. "He did shoot himself."

Paul Meyer takes command of *U-505*. At 7:58 p.m., he records in the official logbook: *Commandant ausggefallen* ("Captain fell out of ranks").

Meanwhile, there's the ongoing crisis of the enemy destroyer above. Meyer deals with it as Zschech could not, ordering the ejection of two *Bold* capsules. These are four-inch-wide metal canisters filled with calcium hydride. When calcium hydride mixes with seawater, huge masses of hydrogen bubbles churn out. The chemical reaction lasts twenty minutes.

The ploy works. The enemy destroyer sends the next round of depth charges in the direction of the bubbles, and *U-505* slips away at the slow speed of two knots. An hour later, she is safely out of range of the destroyer.

Paul Meyer has done well. He has saved *U-505*. At 9:29 p.m., he updates his logbook entry: *Kommandant tot*—"Commanding Officer is dead."

Having shaken off her pursuers, the time has come to inform the crew. No point hiding what everyone already knows. Meyer gets on the intercom.

"First Officer speaking. The captain is dead. We are going to one hundred fifty meters. Silent running. *Das ist alles.*"

So, Zschech is dead. It's official. No details are offered. Meyer is now in command.

Machinist Hans Decker can't believe it. Zschech dead? He thought he heard a pistol shot coming from the control room but couldn't have imagined it was the skipper ending his life. Decker doesn't have time to dwell on it, because just then another round of depth charges detonates. The destroyer has found *U-505* and is in hot pursuit. The attacks continue for two hours before *U-505* can shake off the enemy.

The intercom crackles. It's Meyer again, with a new message.

"To all stations. This is the Exec. We are going home."

Finally, at four in the morning, after the soundman reports no evidence of the destroyer, Meyer orders *U-505* to surface. Just in time, too. The air inside *U-505* is reaching dangerous levels of carbon dioxide contamination. Absorbent granules are spread across the boat to soak up the poison gas.

Oh, how sweet the fresh air tastes.

It is time to bury their skipper at sea.

Zschech's body is placed inside a canvas hammock. A metal plate for weight is wedged between his feet and the hammock is sewn with needle and thread. Goebeler joins in and helps carry Zschech's body out of his cabin, into the control room.

"Control room: Attention!" Meyer shouts. But not one member of the crew stands at attention. There is no salute. Committing suicide in their hour of need is an act of cowardice in the face of the enemy.

Zschech's body is lifted up the ladder to the bridge.

It is just before dawn. The boat is running the surface at top speed. Without any eulogy or words of respect for the fallen leader, the body of twenty-four-year-old Peter Zschech is dumped over the side. Not much of a service, Hans Decker reflects, for Germany's youngest U-boat captain.

Now Paul Meyer has the thorny task of reporting Zschech's death to the German high command. He sends a coded message to Admiral Doenitz's headquarters and to the new Second U-boat Flotilla chief, Kapitän zur See Ernst Kals. Both radio back demanding further information.

How could it be, they want to know, that a German U-boat commander commits suicide in the heat of battle?

THIRTEEN

TURKEYS IN THE AIR

I t takes about ten hours for the USS *Guadalcanal* to cross the Panama
Canal and enter the Caribbean. Everyone on board is on high alert.
These waters are crawling with Nazi U-boats. A U-boat skipper would
earn his nation's acclaim and win the Knight's Cross if he blew up an
American aircraft carrier, the greatest naval target of all. At any given
moment, the *Guadalcanal* could be hit by a torpedo.

After steaming northeast, Gallery swings his ship up the eastern coast
of the United States, past North Carolina's Outer Banks, where more
than a hundred U-boats are believed to be patrolling the waters of the
American defense zone. Cape Hatteras has come to be known as
Torpedo Junction and swimming has been banned, as the ocean is rancid with oil from sunken American tankers.

As *Guadalcanal* moves north to Norfolk, Virginia, Gallery tightens discipline. Even slight infractions of Navy regulations catch his attention.
He posts a handwritten memo addressed to all officers on board. The
subject:

Hands in Pants Pockets
1. *This morning on the flight deck I counted 10 officers with their hands
   in their pants pockets.*
2. *The only reason for which an officer should ever put his hands in his
   pants pockets is to scratch his CENSORED.*

Of course, none of his men dare point out that Gallery wears his cap at a jaunty, nonregulation angle.

Gallery breathes a sigh of relief on December 3, 1943, when *Guadalcanal* docks at Norfolk. He has successfully sailed the *Guadalcanal* from Astoria, Oregon, to Norfolk on the shakedown cruise without, as he puts it, "getting sunk." The crew is "eager-beaver," and Gallery feels good about the *Guadalcanal*—which he now regards as a "reasonable facsimile of a fighting ship." In Gallery's book, that's a rave review.

He leaves his ship in the command of his XO, Jessie Johnson, and chief engineer, Earl Trosino, and heads off to Washington for a round of debriefs with the Navy's division of Anti-Submarine Warfare.

Meeting with the senior officers at ASW, Gallery is issued his orders— he is to take command of hunter-killer Task Group 22.12 consisting of the *Guadalcanal* and four destroyer escorts. They will seek out and destroy enemy submarines operating along convoy routes from the United States to Europe.

There's a lot riding on Task Group 22.12. The *Guadalcanal* is the first baby flattop to be deployed in the Atlantic, and the Pentagon brass is skeptical. If Gallery cracks up too many aircraft, or returns empty-handed with no submarine kills, an entire class of warships, including *Guadalcanal*, could end up as scrap metal. But Gallery is confident that the mass production of baby flattops will alter the course of the war. No longer will U-boats operate with impunity in the mid-Atlantic because they are beyond the range of Allied aircraft. Now American fighter jets and torpedo bombers can swoop down on the subs even a thousand miles from the nearest land base.

Gallery stops by the office of the chief of naval operations and commander in chief of the United States Fleet, Admiral Ernest King. He's known as Uncle Ernie, but never to his face. King is a gruff and irritable warhorse with a Romanesque nose and a deep cleft in his chin. The buzz around Washington is that every morning he shaves with a blowtorch. His daughter once said of him: "He is the most even-tempered man in the Navy. He is always in a rage." When he was named the Navy's top commander, he reportedly remarked, "When the going gets tough, they always send for the sons of bitches." A great line, which King insists he never uttered, but wishes he had. He hates the public relations obligations of his job and would never grant a newspaper interview if he had his way. He would wait for the war to be over and issue a simple two-word communiqué to the American people: "We won."

Fortunately, the meeting goes well, and King has another simple two-word communiqué for Gallery: "Good luck." His unspoken thought may well have been, *You're going to need it.* Off Gallery goes, anxious to leave Navy bureaucracy behind and return to his "sweet ship."

■ ■ ■

New Year's Eve 1943 brings tens of thousands of revelers bursting with optimism to Times Square in New York. Horns blare, noisemakers rattle, and cowbells clang. The taxis are jammed with servicemen on furlough and women dressed in festive evening attire, and when the clock strikes twelve, it's a frenzy of hugging, kissing, and cheering as everyone embraces a joyous moment in these uncertain times.

There is a far more serene scene at St. Patrick's Cathedral and the Cathedral of St. John the Divine, where worshippers bow their heads, praying that 1944 will mark the final year of the world at war.

In Berlin, Hitler issues a grim New Year's message to the German nation, offering further hardship in a war in which he concedes "there will be no victors, but merely survivors and annihilated." The fighting must continue, Hitler declares, with "fanatical hatred." The dictator acknowledges an "apparent slackening of the U-boat war," which he blames on a "single technical invention of our enemies." He does not disclose what that invention might be—but unquestionably it is the installation of radar on Allied aircraft.

The next day, January 2, 1944, *Guadalcanal* sets off from Norfolk on her first wartime cruise, linking up three days later with the USS *Pillsbury* and three other destroyers. In a dispatch, Admiral Royal Ingersoll, commander in chief, Atlantic Fleet, reiterates the mission of Task Group 22.12 in a single sentence: "Operate against enemy submarines in the North Atlantic." That's it. Gallery interprets the order as giving him wide berth. Basically, Ingersoll is telling him, *I have confidence in your judgment—now get going.*

The task group proceeds to Bermuda to top off its fuel tanks, and then on to its designated patrol sector, the Azores.

Gallery doesn't get off to a great start. One of his most experienced pilots is killed coming in for a landing. The plane crashes off the side of the ship and the pilot disappears in a "big splash." Assessing what went wrong, Gallery concludes it was a "little bit his fault, a little bit our fault, and 90 percent bad luck." Gallery doesn't sleep well that night.

The cat-and-mouse game is afoot. Hunting submarines is monotonous work. A U-boat has the upper hand only when she remains invisible. But the submarine's electric storage batteries are only good for about twenty-four hours of continuous operation underwater, at which point the German sub must resurface and run on diesel engines while the batteries recharge.

That is the narrow window in time when Gallery hopes to pounce, when the U-boat is exposed.

As Captain Gallery and his hotshot flyboys hunt Nazi submarines, Wayne Pickels, the coxswain from Texas on board the *Pillsbury*, is engaged in another battle—against rust. Pickels is ordered over the side for painting duty. He keeps a war diary:

January 5, 1944: First day out is rougher than usual. A lot of new seamen are "heaving." Some of the older of the crew are a little sick also. The convoy is slowly forming.

January 6, 1944: Today it was rough again. Rained hard on the morning watch. We got a little painting done this afternoon.

January 7, 1944: Still a trifle rough. We are making a good 10 knots with a good tail wind.

January 8, 1944: Weather calmer today. Had the seamen painting the superstructure, which really need [sic] a good coat of paint. We have a good position in the convoy—8 miles ahead of it. I guess we're the "scout."

January 9, 1944: Painting in the morning. Saw a movie called "Thousands

Cheer" [starring Gene Kelly].

January 10, 1944: Rough weather today. They had hot cakes and couldn't put 'em out fast enough.

January 11, 1944: Seventh day out—did a little painting in the morning but raining in afternoon.

January 12, 1944: Nice day today—got started painting, but too many drills did not help us any.

January 15, 1944: Now headed for the Azores. Writing this on my mid-watch. "Kat," "Doc," and I just shot the bull in the head. Tell them good times we all had in high school. Some times!

It's a touching testament to Pickels's innocence—barely out of high school, and here he is at war, getting excited over hot cakes and watching Gene Kelly dance with a mop in a morale-boosting musical comedy that could have been interrupted by the blast of a Nazi torpedo.

■ ■ ■

It's now two weeks into the mission, and eight of Captain Gallery's "turkeys" from the *Guadalcanal* are in the air, searching for U-boats. So far, they haven't sighted anything. It's looking like another round of aerial patrols about to end in frustration. Just twenty minutes before sunset, it's time to call it a day. The air squadron is ordered to return to the *Guadalcanal*, about forty miles away.

But wait! What's that? Looks like three Nazi submarines on the surface. Two turkeys swing by for a closer look. Luckily for the pilots, it's overcast, so the aircraft approach unseen. One submarine is *U-544*, and she's fatter than the other—she's a *milch* cow, refueling a smaller submarine. The two boats are about a hundred feet apart, connected by a six-inch-thick rubber tether. The third U-boat is farther off, waiting her turn for refueling.

Jackpot! What a perfect setup.

There's nothing the Germans can do to defend themselves. The milk cow is hamstrung by the fuel hose and the towline. She can't crash dive or make any evasive maneuvers. It's like shooting fish in a barrel.

The planes plaster the U-boats with bombs and depth charges. *U-544* is blown to bits. When the plumes subside, the Americans observe flotsam and jetsam floating on the surface, a huge pool of oil, and about thirty Germans swimming for their lives. The second U-boat limps away with the loss of one sailor who has been swept overboard. The third U-boat slinks off and dives, suffering no casualties.

All eight American planes are now circling above. They want a look at the wreckage they have wrought and the spectacle of all those German sailors struggling in the water.

Back on the *Guadalcanal,* cheers erupt when Combat Information Center announces over the loudspeaker that the pilots have taken down one, possibly two, subs. But Gallery is hopping mad. This is no time for rubbernecking. Night is falling, for Chrissakes. He gets on the radio and orders his boys back before it gets too dark, because landing on a baby flattop at night is maybe three times more difficult than landing at night on a regular carrier. There's the shorter deck to deal with, the broad motion of the ship, and the slower speed. And just to make things a little more suspenseful, none of the boys currently in the air has any experience landing on an aircraft carrier in nighttime conditions. It's a skill that only the most highly trained aviators can execute.

The flyboys are so flushed with victory as they circle over the drowning Germans that they fail to notice the sinking sun—that is, until Gallery's angry words send them hightailing it back to the ship.

Meanwhile, the *Guadalcanal* turns into the wind—standard procedure for an aircraft carrier preparing for incoming landings. You don't want to land with the wind at your back on a runway that drops off in just five hundred feet.

Gallery watches from the bridge as the first four planes approach. Aviators on the *Guadalcanal* generally have it pretty sweet. Clean sheets every night. Stewards who make their beds. Steak or stuffed pork chops for dinner. Movies every night. Cheap cigars. It's a real country club—if you live to enjoy it.

The sun has now gone down, and they're all still in the air, burning precious fuel.

The first four planes make safe landings. So far, so good.

But before Gallery can breathe easy, he sees the fifth plane floundering. In the darkness, the pilot can barely make out the carrier deck. He lands too far starboard, and the plane ends up crosswise, with the right wheel skidding into the galley walkway and the left wing and tail sticking out over the deck. The runway is now messed up for the three pilots still in the air.

A crew of sailors tries to physically haul the plane back on deck with pure muscle power. Ten minutes of this fumbling around, and the plane winds up positioned even worse than before. Now the tail is sticking out even farther. Gallery's had enough.

"The hell with it," he barks. "Shove it overboard."

They try everything: prying with 4 x 4 beams, hydraulic jacks, even ramming it with a tractor. They also try cursing, but that doesn't work either. The damn plane just sits there. Gallery has one more trick in his repertoire. He whips the *Guadalcanal* into a tight turn under full rudder and lists his ship 10 degrees to starboard just as the sailors give the plane one more good nudge. That should jettison the plane. Nope, it does not.

Time is running out. Three planes are still waiting their turn to land, and they're practically running on fumes. Via radio, Gallery communicates with his pilots in the air.

"That tail doesn't stick out very far into the landing area," he assures them, not quite convinced himself. "If you land smack on the center line, your right wing will clear it. So, just ignore that plane on the starboard side. Come on in and land."

Darkness *and* a narrower landing strip. Just great.

Here goes.

Without much confidence, the first pilot comes in for a landing.

Gallery grimaces as the pilot makes what looks to him like the widest swipe at a flight deck he's ever seen. The landing signal officer is

Lieutenant Jarvis "Stretch" Jennings, a thirty-one-year-old from Navajo County, Arizona. As you'd expect with a nickname like Stretch, he's well over six feet with uncommonly long limbs. He wears a microphone around his throat and communicates directly with the pilots in a soothing and deliberate voice. He is the law, the ultimate authority for landings. His signals must be obeyed, even if the pilot disagrees. Stretch Jennings waves the aviator off.

Now the next plane gives it a shot. Another wave off from Jennings. About a dozen more attempts follow, but they all get the wave-off.

Finally, a plane approaches "somewhat" near the center of the flight deck. Close enough. They're running out of time. It's do or die. Stretch Jennings gives the signal with his paddles to cut the engines and land. Uh-oh. The pilot is coming in too fast. Here goes nothing. The wheels hit the deck, bounce in the air, and roll over on the plane's back. As if that isn't enough, the pilot then plunges into the sea. What a debacle. Fortunately, the rescue team can fish him out of the ocean.

That does it. No choice now but for Gallery to order the two remaining planes to ditch. He tells his four destroyer escorts to light up. It's a dicey call because they don't know if the German submarine that got away might be lurking about, just itching for vengeance. But the fact is you can't make a safe landing in the water at night without lights, especially in a plane not designed for it. The lights flick on.

Here come the planes, one by one, and the pilots pull it off. Once the crewmen are plucked out of the water, Gallery orders lights off. The task force starts zigzagging the heck out of there just in case that U-boat launches an attack.

That was close. Gallery lost three planes, but at least all the pilots are safe. In the big picture, he's earned one submarine kill, maybe two. Overall assessment: a helluva good day for the Allies.

Now Gallery detaches one of his four destroyer escorts to search for the U-boat survivors forty miles away. It's the law of the sea to make an effort to rescue your enemies after you've sunk their vessel, if you can. By the time the destroyer reaches the site of the encounter with the milk cow, the sea is churning. The odds of finding anyone alive is remote. Several hours later, Gallery calls off the search. He's concerned about leaving his destroyer vulnerable. All fifty-seven German sailors on *U-544* are dead.

Gallery has no regrets. As he points out, he'll have plenty of time to fret about German widows and orphans after the war is won.

That evening, Gallery orders the ship's painter to stencil two miniature swastikas on the bridge to denote two submarine kills, even though officially it's really one.

The USS *Pillsbury* sticks close to *Guadalcanal*. That German U-boat is out there somewhere and could still launch a sneak attack. Wayne "Mack" Pickels is ready to turn in, but he expects general quarters to sound at any moment. A hell of a night, but a lot more exciting than his days painting the ship. As he writes in his diary, "Everyone is sleeping in dungarees with a lifebelt on tonite."

FOURTEEN

A FRAGILE CHARACTER

The ghost of Peter Zschech haunts the room. Since the day of Kapitänleutnant Zschech's suicide, the green curtain to his cabin has never been opened. No one dares enter, and every time Hans Goebeler passes, he gets the jitters. Paul Meyer, now in command of *U-505*, also avoids the cabin. Meyer has every right to sleep in the skipper's bed, but for now he's sticking with his junior officer's bunk.

On November 7, 1943, *U-505* pulls into Lorient. There is no large welcoming party to greet the boat, and no band, not even a harmonica—only a handful of officers and the newly appointed commander of the Second U-boat Flotilla, Kapitän zur See Ernst Kals.

Kals is a thirty-eight-year-old Kriegsmarine hero, the holder of the Knight's Cross of the Iron Cross. In one memorable battle during the Allied invasion of North Africa, Kals slipped his U-boat past an American escort screen and fired off five torpedoes, which hit their targets and sank three US troopships. All in five minutes. That engagement took place on November 12, 1942. Now Kals is acting as Admiral Doenitz's eyes and ears in Lorient.

Kals greets Paul Meyer and praises all hands on *U-505* for their safe return under extraordinary conditions. He doesn't sugarcoat the bleak state of affairs facing the Second U-boat Flotilla. Calamitous losses have hit the U-boat fleet in the last few weeks of the war. The men on *U-505* can see it for themselves. Around the harbor, there are no other submarines docked in any of the pens. The men wonder how many are lying at the bottom of the sea.

Kals orders everyone to maintain silence about the shameful circumstances surrounding Peter Zschech's death. As the first commander in German naval history to take his own life while on active command, Zschech's suicide must be treated as a state secret. The honor of the German officer corps is at stake.

But word is starting to leak out. For one thing, Zschech's bride must be informed, and she apparently tells several close friends what really happened to her husband. Even the Allies seem to know all about Zschech. The night of their return to Lorient, Hans Goebeler and his crewmates are listening to Soldiers' Radio, a British propaganda station operating out of an underground bunker in England. It is forbidden to tune into Soldiers' Radio, but the German sailors listen anyway because they enjoy the station's music and sports programming.

Their ears perk up when a brief announcement is made, specifically directed at the officers of the Second U-boat Flotilla in Lorient. "It must be quite a surprise to you that your friend Peter Zschech did not return with *U-505*."

A typically British tongue-in-cheek jab at the enemy, but the sarcasm isn't what disturbs them. What has Goebeler and the other sailors dumbfounded is the fact that they know about their disgraced captain in the first place.

How could they know? Do the Allies know *everything*?

■ ■ ■

In Berlin, Axel-Olaf Loewe is stunned to hear the news about the suicide of his successor, Peter Zschech. By this time, Loewe is living in a suburb of the German capital with his wife, Helga, their four-year-old daughter, Karin, and their little boy, Axel, age two. Loewe is now serving on Admiral Doenitz's general staff, focusing on armaments and technological development. Hitler has promoted Doenitz to the exalted rank of grand admiral and high commander of the German Kriegsmarine, and Doenitz has brought his protégé, Loewe, with him.

Reading the official accounts of Zschech's suicide, Loewe is taken aback. He always considered Zschech to be a fragile character, but he never suspected something like this would happen. Life for a submarine commander in time of war calls for steady nerves and a strong mental constitution. In Loewe's opinion, Zschech had neither of these qualities. But rotten luck definitely played a role in Zschech's downfall. Analyzing Zschech's record in his single year as skipper of *U-505*, Loewe concludes that Zschech was "consistently pursued" by misfortune. His single accomplishment, such as it was, amounted to the sinking of the freighter *Ocean Justice*, and that was all. Probably a lucky shot, at that.

"Nothing can demoralize a man quicker than lack of success," Loewe says about Zschech. "He lost his nerve. Here a man was asked to give more than he had."

U-505 is now deemed to be jinxed. The most sensible thing to do, in Loewe's opinion, is to scatter the crew to other U-boats in the fleet and start over with fresh personnel. For some reason that Loewe can't figure out, Ernst Kals never follows through with that plan.

In his new staff position, Loewe has vital matters to deal with, namely his growing concern that the entire German U-boat fleet is on the brink of becoming obsolete.

The tide is finally turning in favor of the Allies in the Battle of the Atlantic. By April 1943, the Nazi losses are staggering, with more than ninety U-boats failing to make it back to Lorient. At least half of Doenitz's U-boat fleet lies dead on the ocean floor.

Long-range bombers and aircraft carriers are rolling out of US factories and shipyards on a mass-production scale, and they're hunting and killing U-boats with systematic relentlessness. Countermeasures must be found or the Nazis risk losing the war.

But there's a glimmer of hope for the Nazis, as German scientists invent a nifty new weapon, the acoustic torpedo, codename Falcon. It has an effective range of five thousand meters and comes equipped with a homing system that zeroes in on the sound produced by surface propellers. Even if fired inaccurately, the acoustic torpedo can still hit its target.

There is also the development of the *schnorkel*, a device that enables a submarine to pump in air from the surface while submerged. In theory a U-boat can now stay underwater for weeks at a time, running on diesel, not electricity. Loewe knows of one U-boat commander who remained submerged for fifty-seven days without once surfacing.

But the *schnorkel* has serious technical limitations. For one, the device automatically slams shut in rough seas, which causes an instant reduction in air pressure, and intense pain in the sailors' ears—sometimes rupturing their eardrums. Another flaw is that as the *schnorkel* sucks in air, it vents diesel exhaust. The fumes can be seen escaping on the surface from a distance of almost three miles, alerting enemy ships to the presence of a submerged U-boat.

Loewe is working on other projects. The Elektroboot—a submarine with an extended battery life more than three times the current capacity. And the Walter-Drive, invented by the German engineer Hellmuth Walter—an experimental gas turbine propulsion system that can achieve an extraordinary speed of twenty-eight knots underwater.

The goal, as Loewe lays out for Grand Admiral Doenitz, is to turn the U-boat into a "real underwater vessel"—one that can remain submerged for weeks, even months, at a time.

It's a question of when. If Germany can hold out, fight off the Allies, and wait for technology to catch up, the Nazis will have a chance at victory.

But something is missing from Loewe's analysis. Nobody in the German high command seems to be facing facts and asking the obvious question, which is this:

How the hell do the Allies seem to know the positions of every U-boat at sea?

FIFTEEN

THE OLD MAN

Kapitänleutnant Harald Lange stands before the crew and reads his orders. In doing so, he officially takes over as the third skipper of *U-505*.

At six feet, Lange is the tallest man on *U-505*, and at forty he's the oldest U-boat commander serving on the front lines in the entire German fleet. His voice is a deep baritone that resonates steady leadership and coolness under fire. He is a chain-smoker. There is always a cigarette dangling from his lips, even in the harshest wet weather, although he dares not light up when *U-505* is submerged. That is strictly *verboten*, even for the skipper. His favorite brand is Jan Maat, unfiltered and popular with German sailors because of the image of a sailor on the pack.

It doesn't take long for the crew to start calling their new skipper "the Old Man."

Grand Admiral Doenitz personally selected Harald Lange to be skipper of *U-505*. After the trauma of Peter Zschech's suicide and the acts of sabotage by the French Resistance that made *U-505* the hapless laughingstock of the Second U-boat Flotilla, Doenitz wanted a captain with a proven record of dependability and steadfastness to unite the crew. More a father figure than a rigid disciplinarian.

Lange is also a member of the Nazi Party, number 3,450,040. *U-505*'s first commander, Axel-Olaf Loewe, never joined the Nazi Party. Neither did the despised Peter Zschech. And Lange is no newcomer to the party.

He signed up in the early days, May 1934, a year after Hitler was appointed chancellor.

Lange was born in Hamburg, but he's well acquainted with the United States, having worked as a merchant mariner for the Hamburg America Line in peacetime. He has a first cousin who lives in Indiana. Lange met his German-born wife, a nurse named Carla, in New York City.

At the outbreak of the war, Lange commanded a minesweeper and a patrol boat in the Baltic Sea, then transferred to U-boat service and became first watch officer on *U-180*. Then he graduated U-boat commander school. He is intimately familiar with the waters off West Africa and the Indian Ocean from his days as a merchant marine.

When Axel-Olaf Loewe had his coat of arms painted on the side of the conning tower of *U-505*, he chose a rampaging lion. Zschech scraped that away and replaced it with the Olympic Rings. Now Lange orders his own coat of arms and paints over Zschech's Olympic Rings. In his case, it is the seashell of a scallop. Surely not the most bloodthirsty image to be conjured up in war.

After a long winter's leave, the *U-505* crew is slowly reassembling in Lorient for their next patrol. Machinist Hans Decker had a "disillusioning leave—things were very bad at home." Allied bombing raids over Germany and the collapse of the Eastern Front is making everyone wonder if the Third Reich is doomed. The crewmen who visited with their families in Berlin or Frankfurt return to Lorient with horror stories about having spent most of their time hiding in underground air raid shelters. Even small towns are getting pounded. The railroad station near Marburg, close to Hans Goebeler's village, is wiped out of existence during an aerial attack. But for true-believer Goebeler, the Fatherland coming under aerial bombardment only serves to make him "fight even harder."

As provisions are loaded, Harald Lange warns his sailors to keep a sharp eye on the French dockworkers. Sabotage will not be tolerated. The lesson learned from Zschech's experience has been drummed into the new skipper: *Don't trust the French.*

The same goes for drinking and whores in Lorient. They are "bad influences." Contracting a venereal disease is a serious offense, punishable by court martial and three months in prison. Of course, none of this stops the sailors from heading over to the local bordellos for a quick one. Harald Lange turns a blind eye. He knows the way of sailors from his long career at sea.

Christmas Eve 1943. A big blowout party for the crew of *U-505*—a twelve-hour bacchanal of feasting and boozing. At midnight, they all conk out. The next day, hung over and rubbing the sleep out of their eyes, they set out for the first time under Harald Lange's command. It's strange to see him at the controls instead of Loewe or Zschech, but the crew seems to agree that Lange is the right skipper for their boat. Paul Meyer, who took over after Zschech's suicide and earned praise for bringing *U-505* safely back to Lorient, remains second in command. Considering his outstanding performance, Meyer expected to be named skipper. Despite his disappointment, Meyer makes it his goal to get along with Lange.

Three days out, they hear the faint sound of artillery shells and bombs exploding somewhere in the distance. A major sea battle is taking place about sixty miles away, pitting five German destroyers and a squadron of scrappy T-12 German torpedo boats against a force of British warships. The Battle of the Bay of Biscay is underway.

Flotilla headquarters radios new orders. *U-505* is to proceed at top speed and search for survivors of the German destroyer *Z-27*, which has been hit by shells fired from twelve miles away by the British light cruiser HMS *Glasgow*. The destroyer is dead in the water, and *Glasgow*, in hot pursuit, opens fire at point-blank range for the finishing touches, sinking her and killing about three hundred sailors. Another ninety-three Germans are in the water, awaiting rescue. It's a big victory for the British.

Lange takes to the bridge, lighting one soggy Jan Maat cigarette after another as his face is battered by ten-foot swells and the bone-chilling cold. From the control room below, Hans Goebeler catches a glimpse of Lange soaking wet. He's impressed. Here at last is a commander worthy of respect. Every blanket is gathered in preparation for the shipwrecked *Z-27* sailors, should any be found. Toni, the cook, has all the burners firing to keep the coffee coming.

A red flare! *U-505* swings around and heads in its direction. It takes another two hours before they spot the source: two German sailors clinging to a one-man life raft. The men are fished out of the water and taken below deck where they are stripped of their soaked clothes, rubbed dry, and carried into the engine room to warm up.

"Damned Tommies!" Lange sputters, uttering the slang for the British military.

Three hours later, a glimpse of something riding the crest of a big wave. *U-505* maneuvers closer. Now they see it. It's a cluster of rafts lashed together, with twenty-seven shivering German sailors in woeful condition. A towline is tossed, and one by one the men are lifted aboard *U-505*. They are delirious with hypothermia. It turns out the sailors aren't from the destroyer *Z-27* after all. They're the crew of a T-25 torpedo boat sunk during the battle.

In all, thirty-four shipwrecked sailors from the Battle of the Bay of Biscay are rescued by *U-505*. Finally, Lange gives up, deciding the wisest course of action is to slip out of there before his vessel is spotted by the British. He turns *U-505* around and heads back to France to drop off the survivors. Not to Lorient this time, but to the port city of Brest, headquarters of the First and Ninth U-boat Flotillas.

They pull into Brest on New Year's Day 1944. Before they disembark, the T-25 sailors sign a guest book expressing appreciation:

[A]nd we are thankful that a kind fate will enable us to start very soon on another mission against the enemy. Let us hope that we shall soon succeed in forcing him to his knees. In this hope we wish the U-505 *lots of luck in the safe return.*

SIXTEEN

LONE WOLF

Thirteen Months Earlier
On Board *U-515*, Sister Ship of *U-505*

W hat a beautiful ship. Pity she must be destroyed.

This is what Kapitänleutnant Werner Henke is thinking as he stands on the bridge of *U-515* in the South Atlantic, west of the Azores, admiring the British ocean liner SS *Ceramic* through his binoculars.

The date is December 6, 1942, thirteen months before *U-505*'s rescue of thirty-four German sailors shipwrecked in the Battle of the Bay of Biscay.

On this night, *U-505*'s sister ship—*U-515*—is riding the surface twelve hundred meters (about three-quarters of a mile) off the SS *Ceramic*'s starboard beam. Henke estimates the ship's weight at just under twenty thousand tons. Quite satisfactory, Henke decides.

For the ambitious U-boat commander, this could be his first "really big hit" of the war, and he is determined to take her down. The U-boat skipper is thirty-three, handsome, and blond, with penetrating blue eyes. At five-foot-nine and 175 pounds, Henke could be a living embodiment of Hitler's concept for the master race. He is fond of American jazz, dancing, and Wagnerian opera. His idol in German history is the nineteenth-century Prussian statesman Otto von Bismarck.

As night falls, *Ceramic* kicks up her speed and breaks into a zigzagging course. All lights are turned off. Evidently, the British captain is apprehensive about the reported presence of U-boats in these waters and orders blackout conditions.

Henke conducts an end-around maneuver.

At 8:00 p.m., he is ready. "Tubes One and Four, prepare for a surface firing!"

The torpedo officer responds, "Flooding Tubes One and Four, opening outer caps. Tubes One and Four ready."

"Depth five meters. Torpedo speed, forty. Target sweet 15.5 knots. Range twelve hundred meters."

"All set!"

Henke calls out, *Facher, los!*—"Salvo, fire!"

Thirty seconds later, a muffled explosion.

The torpedo hits the *Ceramic* in the forward section. A second torpedo also scores but hits the steel hulk with a clink of metal on metal. It's a dud.

Just the same, *Ceramic* is dead in the water, although she doesn't show any sign she's ready to sink. Henke swings his U-boat around and fires another torpedo at a range of one thousand meters. It's a direct hit. Henke observes the chaos as lifeboats are lowered and passengers and crew abandon ship. *Ceramic* now lies deep in the water. Henke shoots one more eel to "hasten her demise." It's the coup de grâce. What happens next reminds Henke of a "mighty fist from below" that breaks the great ship in half. The masts collapse. He has never seen anything as terrible as this. In six seconds, *Ceramic* is gone.

Suddenly, an underwater explosion. It could be the ship's boilers, or perhaps ammunition stored inside her cargo holds. Henke turns *U-515* and heads northwest, reporting the sinking of the *Ceramic* to Admiral Doenitz and U-boat command headquarters. He's expecting congratulations, and is stunned by the frosty response that comes in at seven thirty the following morning:

REPORT AT ONCE WHETHER TROOP TRANSPORT WAS LOADED WITH TROOPS AND WHETHER THERE WAS ANY INDICATION OF ITS PORT OF DESTINATION.

Reading the radio dispatch, Henke is vexed. Is Doenitz serious? The only way to confirm if the *Ceramic* was transporting Allied troops is to return to the site of the sinking, which by now must certainly be crawling with enemy destroyers searching for survivors. In any event, the passengers wore British military uniforms. He saw them! What more proof is necessary? Henke lets his irritation simmer for a while, then he proceeds to follow orders. Nothing he can do now but turn *U-515* around and head back. What a waste of fuel. He sends a dispatch dripping with sarcasm: AM GOING TO THE SITE OF SINKING IN ORDER TO CAPTURE THE CAPTAIN.

He proceeds to retrace his submarine's path, but behind his bluster Henke is worried. In wartime, heroism and disgrace are often separated by a thin membrane. He wonders if he may have pierced it by sinking the *Ceramic* and killing everyone aboard.

Well, not quite everyone.

■ ■ ■

Eric Munday is a sapper in the Royal Engineers. In the British military, a sapper is a trained combat soldier who specializes in demolitions, bridge-building, and breaching fortifications. Sappers also repair roads and airfields. Before the war, Munday worked as a clerk on the Southern Railway in England and played on the company's football and cricket teams.

At 8:00 p.m. on the night of December 6, Munday is in the smoke room of the SS *Ceramic* playing the card game English Solo with three Army blokes—Andy, Harry, and Jack. The *Ceramic* is a luxury ship with a notable history. She is often compared to the *Titanic* in that she was built in 1913, one year after the *Titanic* disaster, also in Belfast. She actually resembles that legendary ship, gone now thirty years. Like the *Titanic*, she's a White Star Line ocean vessel. In peacetime, her normal run was transporting passengers between Australia and England, but in 1940 she was requisitioned by the UK Ministry of Transport for the duration of the war. All her lower passenger cabins were converted into troop quarters.

Munday boarded *Ceramic* in Liverpool on November 23, the day the ship set sail. The 641 passengers included thirty nurses and twelve children, most of them dependents of British military personnel who were also on the ship. The youngest was a one-year-old girl. The ship was en route to Sydney, Australia, by way of refueling in Saint Helena Island in the South Atlantic Ocean and Cape Town, South Africa.

It has been a pleasant voyage thus far. You wouldn't know there's a war going on based on the food they serve. Kippers and eggs for breakfast, or bacon and eggs—your choice. Lovely nurses to flirt with in the bar. Munday considers himself lucky to be on board, until the first torpedo hits.

There are forty-five survivors in Munday's lifeboat. By midnight, the weather worsens, with terrifying downpours and high winds. Huge waves are breaking over the lifeboat. Munday rows all night, and when he's not rowing he's bailing. By morning, a northerly gale has moved in. Munday doesn't know if he has the strength to make it another day. He thinks about his mum and dad back in Surrey and a girl named Pam—a shame they didn't get to know each other better before he set out for war.

Munday is sitting forward on the starboard side when the lifeboat suddenly capsizes. He flops into the water, with the other lads falling on top of him. He's going to drown for sure in this tangle of flailing limbs, but no—he sees a gap and swims to the surface, gasping for air. After a struggle, the men right the boat but it's three-quarters filled with water, and it's Munday's opinion the lifeboat is no longer seaworthy. All he can do is hang on to the side. Then he makes up his mind to take his chances elsewhere and swims off on his own, looking for another boat or maybe a chunk of floating debris he can hang on to. He's always been a strong swimmer. He's lucky to be wearing a life jacket.

Four agonizing hours pass. Munday doesn't think he can last much longer.

What's that?

In the violent storm, it's hard to make out anything, but Munday swears there's an enormous shape in the water just a hundred yards away. He starts swimming in that direction. There it is! Dear God, it's a submarine. A giant swell carries him closer. Now he can make out several sailors on the conning tower. It's a German sub. Never did he think he'd be grateful for the appearance of an enemy U-boat, but there it is. What a blessing, unless they open fire on him.

They toss him a rope, and Munday hangs on tight just as another wave carries him to the deck of the U-boat.

Munday is taken below, finding himself in the remarkable position of being held a British prisoner inside a German U-boat. He's swallowed so much seawater the Germans have to press down on his stomach to expel the water. Then he's given dry clothes, condensed milk, and a hammock to rest in.

Werner Henke interrogates the prisoner. Munday is hardly the sunken ship's captain that Henke vowed to capture, but he'll do. Munday's first response is to declare, "I won't talk." He claims not to know the *Ceramic*'s cargo or destination but confirms the presence of British troops on board.

Henke hears enough. As far as he's concerned, the *Ceramic* was a troop transport and therefore a legitimate target in time of war.

Munday is the sole survivor of the *Ceramic* disaster. No one else from the ship is found alive—they're all gone, 656 passengers and crew. All the nurses and children and the soldiers, killed by the torpedo blasts or drowned in the raging storm.

And Munday must wonder: *Why me?* Munday starts a diary to try to make sense of it all.

8th Dec: I awoke feeling very sore and aching so much I could hardly move. Everything was very strange and besides feeling none too good I felt very depressed.

10th Dec: Today they gave me a tooth brush, paste and a comb. They let me have my first breath of fresh air and to me it was like a million dollars. I was also allowed to smoke a cigaret.

12th Dec: My 21st birthday, but one vastly different to one I had meant to have.

13th Dec: Life in a U-Boat is no joy ride. It is just like riding in the underground but much worse.

On December 14, *U-515* happens across an unusual object floating in the ocean—fifty cases of New Zealand butter. It's booty from the British merchant ship *Hororata*, sunk just a day ago by *U-130*.

23rd Dec: They have started to prepare the food for Christmas. As they have plenty of butter they are making piles of cream to pour over the cakes.

24th Dec: All the lights were turned off and they lit a small Christmas tree; it was very effective. Then they sang songs, equivalent to our carols I believe, because one of them was to the tune of our carol, "Silent Night." We drank hot punch and ate chocolate and biscuits.

There are no idle hands on board a submarine. Henke puts Munday to work peeling potatoes. Munday develops a boil on his neck the "size of a football." The ship's doctor says it's the largest boil he has ever seen.

The Germans come up with a nickname for Munday—"Johnny." They take a liking to him. As *U-515* nears Lorient, the crew advises Munday to stuff his face because "once you get into Germany you won't get any food like this."

On January 6, *U-515* pulls into Lorient, greeted by a band, military honor guard, and a contingent from the Women's Naval Auxiliary. The flotilla commander boards the boat and presents Warner Henke with the Knight's Cross of the Iron Cross. Henke's sinking of the SS *Ceramic* is about to make him a national hero in Germany.

Armed guards take Eric Munday down the pier to the arsenal for interrogation. A German radio reporter wearing the uniform of a naval officer unexpectedly thrusts a microphone into his face. Being in no position to refuse, Munday answers a few questions. All he wants to do is let his mum and dad know he is alive.

REPORTER: "You were the only survivor of the *Ceramic*?"

MUNDAY: "Yes, I think so, that I am the only survivor of the ship. It was very bad."

Radio Berlin also interviews Werner Henke about the sinking of the *Ceramic*.

REPORTER: "So, she was a troop transport?"

HENKE: "Whether she was a dedicated troop transport, I couldn't say. However, she had lots of troops on board. On the following day I returned once more to the site of the sinking, and I found a most horrific site. I noted many corpses. It was not a very nice sight. I tried to help several people. However, with the sea being that rough, and gale force winds, it was not really possible to carry out rescue missions. I managed to rescue an English soldier by pulling him aboard. I nearly lost my first officer and a crewman during the maneuver, and I finally broke off. In any case, my job is not to rescue people but to wage war on the enemy."

In Thorton Heath in South London, traveling salesman Ernest Munday and his wife are awakened by a telephone call from a stranger who is listening to the Radio Berlin broadcast.

"Have you a son called Eric? Because I have just heard his name mentioned on the radio."

Pretty soon, a correspondent for the London *Standard* newspaper comes knocking on their door. They are overjoyed at the news that their son is alive.

"I can hardly believe it," says Mrs. Munday.

While Germany hails Werner Henke as a hero, the Office of War Information in Washington counterpunches with a propaganda radio report that accuses Henke of mowing down the shipwrecked survivors of the *Ceramic* with machine-gun fire, as they sat in their lifeboats waiting to be rescued.

War is war. But there is a tremendous difference between war and murder.
Henke had the water combed with his searchlights and whenever he found a boat with survivors, or women and children . . . clad in life-saving

jackets, struggling with the waves—he had his guns trained up on them. Yes, he ordered the survivors slaughtered with machine guns. He drove his crew to supreme efforts in order to murder as many of the helpless people as possible.

The broadcast goes on to assert that Henke is a war criminal.

Nothing will be forgotten. And one day, when there is peace again, a German court—not a Nazi court—will condemn these criminals. Among the men who will face the judges there will also be Lieutenant Werner Henke.

Only Eric Munday could bear witness to what really happened the night the *Ceramic* was sunk—but he's now a POW in Germany in Stalag 344 and incommunicado.

Werner Henke becomes a famous U-boat skipper in Germany. The highest honor is awarded when Henke is summoned to the Wolf's Lair, Hitler's military fortress in East Prussia. This is where the Führer is guarded by SS units and vigilant food tasters on alert for poison. On July 4, 1943—America's Independence Day—Henke stands at attention before Hitler as the dictator hands him a black-hinged case containing the Oak Leaves clasp for his Knight's Cross.

Of course, Henke has no way of knowing how his fame for having torpedoed *Ceramic* will one day come back to haunt him—and his sister boat, *U-505*.

SEVENTEEN

PING . . . PING . . . PING

Eight Months Later
April 8, 1944

It's a German U-boat all right.

Lieutenant Commander Richard Gould can barely believe his eyes. He's flying his stubby, single-seat Grumman F4F Wildcat right over the U-boat, and he's looking straight into the conning tower hatch. He can see the lights shining inside! He circles to make a fresh run, but by the time he swings around the submarine is gone. Must have crash dived. *Damn!*

An hour later, Gould is back on the USS *Guadalcanal*. Gould is a twenty-nine-year-old Annapolis grad who earned his wings in 1941. He hails from Washington State where his father sells ads for the local newspaper and his mother teaches community college. Waiting for him back home is Mary, his high school sweetheart and wife of three years.

"Cap'n," Gould says breathlessly. "I almost got him!"

Captain Dan Gallery does a double take. He's in the Combat Information Center debriefing the pilots just back from another uneventful air patrol and stares at Gould with a puzzled expression.

"That submarine," Gould explains. "He barely got under in time to get away."

This is the first Gallery is hearing about a confirmed sighting of a submarine. He's boiling mad. Why wasn't he informed? Apparently, there was a miscommunication. Gould says he radioed the sub sighting and got a "Roger" back in confirmation, but for some reason the report never reached CIC. Gallery says he'll deal with the screwup later. Right now, he wants all the dope on that sub.

"Are you sure it was a submarine?" Gallery asks.

"Absolutely certain. I looked right down the conning tower."

"How certain are you of the position?"

"If you'll give me about two minutes to work it out, I'll give you a real good position on it."

"Okay, get busy."

It's now 0130 hours—the dead of night.

The first hours of Easter Sunday.

■ ■ ■

The *Guadalcanal* set out from Norfolk a month ago, on March 7, 1944. Task Group 21.12 was escorted by the destroyer USS *Forrest* and four destroyer escorts—*Pillsbury*, *Chatelain*, *Pope*, and *Flaherty*. Hunting grounds were the waters between Gibraltar and the Azores, in a shipping lane swarming with U-boats emerging out of the Bay of Biscay.

Their first day out, Gallery got the pilots together for a preflight briefing in the *Guadalcanal* ready room and explained what he had in mind. They didn't have to be reminded about what happened three months ago when Gallery's planes were forced to ditch in the ocean because his pilots were not trained in night landing. Now Gallery is telling the lads he is determined not to be caught up in that kind of sticky situation again. Only he uses more choice words.

No baby flattop captain in any theater of war had yet to authorize nighttime flight operations off an aircraft carrier. Full-size carriers, yes. But not the baby flattops. Gallery will be the first. Great if it works—he'll be hailed as an innovative genius. If he fails, he risks losing a lot of planes and pilots—specifically the men in the ready room who are staring at Gallery as he lays it all out.

Gallery waited for the appearance of a full moon rising before commencing training—ideal conditions for night operations. It was a thing of beauty, watching the planes take off as the light from the heavenly body flooded the deck of the *Guadalcanal* with abundant illumination. Each night, as the moon waned and the skies darkened, the pilots became more confident in their nighttime flying capabilities. In two weeks' time, the moon disappeared and they were taking off and landing in pitch-black conditions. Dangerous work. A third of the planes on the *Guadalcanal* were wrecked. Fortunately, no pilot was lost—they all managed to scramble out, shaken but unscathed.

So now, only two-thirds of his planes are deemed airworthy, but Gallery still believes the losses are worth it. Fewer aircraft, maybe so, but twice as effective is how he figures, because his fleet now has the capacity to fly around-the-clock, not just in daylight, when subs are hiding underwater. Gallery believes this tactic could break the back of the U-boat menace. Until now, German submarine skippers operated pretty much without fear of a nighttime aerial attack, when they could recharge their batteries with impunity and hunt for prey. That's done with. Henceforth, the U-boats surface at night at their own peril. Only they don't know it yet.

■ ■ ■

Fast-forward to Saturday, April 8. The *Guadalcanal* is on the thirty-third day of this mission, not having had much luck hunting submarines. Task Group 21.12 has refueled in Casablanca and is now on its way back to Norfolk without a single kill. Gallery has a squadron of Grumman Wildcats and Avenger torpedo bombers in the air, in a sweep extending sixty miles in front of and one hundred miles on either side of the *Guadalcanal*. This is when, at 10:00 p.m., Saturday, April 8, Lieutenant Commander Dick Gould looks into the open mouth of the German conning tower hatch.

It's now dawn on Easter Sunday.

Four "turkeys" have taken off from *Guadalcanal*, en route to the U-boat's last reported position based on Gould's sighting.

There's the sub! She's about fifteen miles from where Gould first laid eyes on her.

A fighter pilot takes aim and drops two MK-47 depth charges. The U-boat throws up heavy antiaircraft fire. The battle is joined.

Back on the *Guadalcanal*, Gallery is monitoring the engagement via radio communication, eager for good news.

"All depth charges fell short," the pilot reports. "Estimate no damage."

Then, more disappointing news from the pilot: "Sub has submerged again."

But it's not over. The pilot can see the U-boat silhouette. She's trying to surface and recharge—the German skipper must be desperate. Then the U-boat hotfoots it out of there when the skipper realizes the Yank plane is flying right over, about to pounce. Gallery nods. The noose is tightening. At least they've got a solid new fix. Gallery figures the U-boat could not have fully recharged her batteries in the limited time she was on the surface before crash diving. He figures that sub is a dead duck by tomorrow.

Gallery orders three destroyer escorts to peel away from *Guadalcanal* and smoke out the enemy boat with sonar.

One hour after sunrise, they detect something.

The *Pope* drops a salvo of depth charges. The ash cans arch into the air, hit the water, and sink to the specified depth. Seconds later, *boom!*

Two more attacks follow. A dozen depth charges are dropped in an expanding pattern.

Then quiet. The sonar operator must wait for the ocean to settle before he can regain contact. Otherwise, he acquires false readings.

Ping . . . Ping . . . Ping.

The U-boat is still alive.

More ash cans. The next moment, a trickle of garbage and oil gurgles to the surface.

Nice try, Gallery thinks. He's not falling for that old trick. By this stage of the war, it's a predictable U-boat ruse—pumping debris to the surface to deceive the Allies into thinking the depth charges have hit their target and the submarine has been destroyed.

Lo and behold, moments later . . . *Ping . . . Ping . . . Ping.*

Crafty bastard, Gallery reckons. More proof that the U-boat skipper, whoever he is, is a "tough customer who knows his business."

Judging from the sonar bounce, the sonar operator figures the U-boat is running deep at five hundred feet, fleeing for her life.

Actually, she's now at six hundred feet—maximum dive range.

At 2:10 p.m., nineteen depth charges fired by *Pope* finally delivers the dagger. It's not a direct hit, but the U-boat is left battered by hydraulic shock waves. The hull is breached. Water comes flooding in. It's over.

Inside the U-boat, the skipper shouts: "Blow all tanks. Prepare to abandon ship and scuttle!"

Seven minutes pass before the U-boat breaks surface—and finds herself surrounded by four American destroyers: *Pillsbury*, *Pope*, *Chatelain*, and *Flaherty*. They're less than a hundred yards away, ready to open fire. Gallery is a good five miles away, on the bridge of the *Gaudalcanal*, following the clash via radio. He'd certainly like to be there right on the scene, but that's not how aircraft carriers operate. They let the destroyers do the fighting. The carriers exist as floating runways for takeoffs and landings, and the destroyers exist to shield the carriers. Tactically, the aircraft carrier is required to turn tail and run as soon as contact is made with an enemy sub. America does not want to lose an aircraft carrier.

As the U-boat breaks surface, there are just seconds to assess the intentions of the German skipper. No one knows for sure what will happen next. Strange, but the U-boat looks structurally sound. Did she surface with the objective of taking down as many ships as she can? Is the U-boat skipper bent on departing this earth in a blaze of glory for the Fatherland? Will she open fire with her torpedoes? Fire away with her cannons? Or will the skipper surrender without a fight?

On board the *Pillsbury*, Boatswain Mate Second Class Wayne Pickels is watching the battle play out from the bridge and is startled to realize a torpedo is racing right for his ship. He can see its wake.

"Torpedo coming!" he shouts.

So the German skipper has apparently decided to go down fighting. What balls!

The commander of the *Pillsbury*, Lieutenant Commander Francis L. Dale, is a prominent Cincinnati lawyer in civilian life. That's good, because lawyers know the value of evasive maneuvers—and this sure is time for one of those.

"Flank speed!" Dale shouts. The *Pillsbury* zigzags, and the torpedo whizzes by. But Pickels can't believe what happens next. The German torpedo is turning! That eel is *following* the *Pillsbury*. The damn thing must be that new German invention they've been warned about—the acoustical torpedo. And it's heading right for the *Pillsbury*'s propellers.

The chief boatswain hollers out, "Tell the skipper here comes one of those sons-of-bitches with ears!"

Then, the torpedo just vanishes, having expended its fuel supply and sinking without making contact.

Well, that settles it. The intentions of the U-boat skipper are now obvious.

The Americans let her have it.

All four destroyers open up with everything in their arsenals: depth charges, torpedoes, four-inch guns, .20mm antiaircraft guns.

In the air, flying his F4F Wildcat, Dick Gould presses the attack, strafing the U-boat with six .50-caliber, wing-mounted machine guns.

German sailors start popping out of the fore and aft escape hatches and dive over the side. The bow of the stricken U-boat slowly rears straight into the air, her stern down. A rush of water is pouring out of the vent holes.

"Thar she blows," cries XO Jessie Johnson, standing at Gallery's side on the *Guadalcanal*. The U-boat looks like a stricken whale. Yes, she's definitely a goner. There'll be no further acoustic torpedoes fired from that boat.

Dick Gould swoops and drops a life raft to the Germans who are struggling to stay afloat—a humanitarian gesture in the heat of battle. (Gould will later be awarded the Distinguished Flying Cross for displaying "cool courage and expert airmanship.")

The process of picking up survivors gets underway. *Pope* takes primary responsibility for plucking the Germans out of the water because she is the nearest ship. The skipper of the *Pope* is Lieutenant Commander Harvey Headland, thirty-three, an Annapolis grad and the son of a North Dakota banker who lost his business during the Great Depression. On December 6, 1941, he was attending a dance at Mills College in Oakland, California, with a pretty young student from Tacoma named

Margaret McGinnis. He heard himself tell Margaret, "I love you," and asked her to marry him.

Quickest engagement ever, because the next day the Japanese bombed Pearl Harbor, which was still smoldering when Margaret and Harvey exchanged "I dos." They kissed each other goodbye—she went back to college, he reported for sea duty, and now here he is three years later, commander of the USS *Pope*, looking on a scene more hellish than any he could have imagined.

German sailors—their heads bobbing up and down—scream for help as they struggle to stay afloat. Some take their last gasps while being swallowed by the ocean, their outstretched hands vanishing last, never to be seen again.

One by one those who were able to stay afloat are plucked out of the water and brought aboard the *Pope*. Sixteen German sailors have lost their lives. In all, more than three dozen sailors are rescued. They look dazed. Just about every German suffered punctured eardrums due to the depth charges that battered their boat with shock waves.

One German climbs aboard and surprises everyone when he declares in English and in the most agreeable tone, "Happy Easter!" He seems enormously thankful to be alive.

He and the other prisoners are escorted under guard to the wardroom for interrogation, the officers first.

Only now do the Americans learn the identity of the boat they just sank. She is *U-515*, skippered by Werner Henke.

Headland leaves the bridge. He is curious to check out these POWs.

In the wardroom, he encounters Henke. Curious about his state of mind, Headland asks Henke who will win the war.

Why, Germany, of course, Henke answers. No doubt about it.

Headland asks when this remarkable event will occur.

Henke's reply: "1944."

Another officer from *U-515* undergoes questioning. Are there any other U-boats prowling about, he's asked. Suddenly, the prisoner stands ramrod straight and his arm shoots up in the Nazi salute.

"Heil Hitler!"

Headland sneers in disgust. He really wants to kick the Nazi in the rear end but thinks better of it.

The POWs are transported by whaleboat to the *Guadalcanal*, which has the facilities to detain them under lock and key. When they are safely on board the carrier, Gallery orders his chief master-at-arms to bring the skipper of *U-515* to his cabin. For the time being, the identity of the boat, *U-515*, is just a number to Gallery. It means nothing to him.

Pretty soon, a husky, good-looking blond officer is brought before Gallery. It's strange to see this German wearing GI-issued dungarees and a sweatshirt.

"This is the captain, sir," announces the master-at-arms.

No need to say it. Even in captivity and stripped of his uniform, the prisoner has a brash bearing. Seems like a pretty arrogant guy for someone whose ship was just blown to smithereens.

"Your name?"

"Henke. Werner Henke, *kapitänleutnant*, Kriegsmarine." Henke recites his serial number. His English is impeccable.

"The number of your U-boat?"

Henke doesn't answer.

Of course, Gallery already has that information. "It *was U-515*," he says. "You sailed from Lorient ten days ago." Gallery is playing mind games. He doesn't really know with certainty *U-515*'s track, but he wants Henke to think he knows more than he does. It takes a U-boat nine days or so to sneak out of the Bay of Biscay. So, Gallery doesn't think he's far off when he says ten days.

Henke shrugs, not confirming or denying.

Gallery starts to ask another question. "How many ships—"

"Captain, I have a protest to make!"

"What is it?"

"You violated international law and the Geneva Conventions."

"How?"

"You killed many of my men while we were trying to surrender."

Gallery swats that one away. "I had no way of knowing whether you were trying to surrender or to torpedo my ships. As soon as we were sure you were harmless, we ceased firing and we have rescued forty-five of your men."

"But you killed ten [sic]—in violation of the Geneva Conventions."

Gallery shakes his head. No point in getting into a debate with a Nazi. He turns to his master-at-arms. "Take him below."

Sometimes, as a gesture of kinship with a fellow commander, an American skipper will offer a captured U-boat captain his cabin. No way with this nasty character, Gallery decides. Not a chance. Henke is taken to the brig, where the other officers of *U-515* are locked up.

The next day, the master-at-arms sends word to Gallery that Henke is demanding to see him again.

"What for?" Gallery asks.

"Something about the Geneva Conventions."

"Oh, that again. Okay."

When Henke stands before Gallery, he starts rattling off a litany of complaints about being incarcerated in the brig with the other officers. According to the Geneva Conventions (at least in Henke's understanding), he is entitled to an officer's stateroom and meals in the officers' mess.

Gallery explains that he doesn't happen to have a copy of the Geneva Conventions handy but he isn't about to evict one of his officers from his bunk for a Nazi regardless of what the Geneva Conventions may or may not say.

"Besides," he says, "many of my officers and sailors are of Jewish or Polish ancestry. They might not be very polite to you if I gave you freedom of the ship."

"According to the Geneva Conventions, it is your duty to protect your prisoners."

Gallery rolls his eyes. He's just about had it.

"Captain, we are going to refuel in Gibraltar about ten days from now. If you don't like the way I'm treating you, I'll be glad to turn you and your crew over to the British. Maybe they will treat you better."

This triggers a strange reaction from Henke. All the bluster is suddenly sucked out of him. He stares at Gallery. Evidently, he is reconsidering his position. "I can put up with this treatment for a few more weeks," he says. "I withdraw my protest."

Really? Just like that? With that, Henke is escorted back to the brig.

But the back-and-forth gets Gallery thinking. Something weird has just taken place. Henke went from belligerent to meek in the blink of an eye. Gallery's master-at-arms saw it, too. He smells a rat and does some digging and learns a few interesting facts about Werner Henke from the other prisoners.

First, Henke turns out to be quite a catch—he's one of Grand Admiral Doenitz's U-boat aces and a recipient of the Knight's Cross of the Iron Cross. Second, his crew "hates his guts." They think Henke is a reckless commander who takes too many risks because he wants to win Oak Leaves with diamond studs to add to his Knight's Cross. But the most stunning morsel of information is this—Henke is wanted for war crimes by the British. The accusation is monstrous—that he sank the HMS *Ceramic* in December 1942 and machine-gunned the ship's survivors as they waited for rescue.

Gallery ponders what he's hearing. He realizes the war crimes accusation could be the reason Henke backed off so abruptly when Gallery threatened to transfer him to British custody in Gibraltar. The wheels start churning in Gallery's head. Time for a "shenanigan." He calls his XO, Jessie Johnson, and fills him in on the little caper he's about to carry out. He has the boys in communications type out a phony dispatch on blank paper. Then they send for Henke.

Gallery hands Henke the official-looking PRIORITY dispatch, from the commander in chief, Atlantic Fleet.

Henke reads:

BRITISH ADMIRALTY REQUESTS HENKE AND CREW OF *U-515* BE TURNED OVER TO THEM IN VIEW OF CROWDED CONDITION OF YOUR SHIP YOU ARE AUTHORIZED TO USE YOUR DISCRETION.

Henke's eyes widen. Looks like he bought it.

"Why do they want me?"

"I don't know," Gallery answers.

Henke pauses. If he's turned over to the Brits, he is certain he will be tried and convicted for war crimes and hung, even though there is no proof he slaughtered the *Ceramic* survivors in their lifeboats.

"Well, Captain, I suppose there is nothing you can do about it."

"Yes, there is," Gallery says. "That dispatch allows me to use my discretion. If you make it worth my while, I'll keep you on board until we arrive in the US."

"What do you want me to do?"

"Just sign this." Gallery reaches for a prepared statement and pushes it across his desk. It's an official document, typed out on legal paper, with the seal of the USS *Guadalcanal* embossed on it.

Henke reads:

I, Captain Lieutenant Werner Henke, promise on my honor as a German officer that if I and my crew are imprisoned in the United States instead of in England, I will answer all questions truthfully when I am interrogated by Naval Intelligence Officers.

Signed
Kapt, Lt.

Witness:

D. V. Gallery, Capt, USN

J. S. Johnson, Cdr, USN

Henke looks up. The document requires him—a hero of the Third Reich personally decorated by Hitler—to divulge his country's military secrets.

"Captain, you know I can't sign that."

"It's up to you. Sign and you go to the United States. If you don't sign then you and your crew go to England."

Henke reads it over two more times. He's in a real squeeze. Go to England and face execution even if the war crimes allegation is manufactured or proceed to the United States and live—but first he must betray the Fatherland. Of course, it's all a colossal bluff by Gallery, whose intuition tells him that Henke is ready to crack.

Henke sighs, long and hard. "Well, Captain, what would you do if you were in my position?"

Gallery responds, "If I were convinced that my country had lost the war and that I could help my crew by signing—I would sign."

Henke picks up a pen. Just twenty-three days ago, while on leave, Henke married a widow he met during a skiing vacation. He has every incentive to stay alive and to look forward to a happy life with his wealthy bride after the war. He signs the document.

With Henke's signature in hand, Gallery orders the POW to be returned to the brig. In quick order, Gallery makes photostat copies of Henke's agreement and has them circulated to the imprisoned crew of *U-515*. Of course, they all recognize Henke's signature and reach the same calculation. As Gallery puts it, "The skipper is talking—why shouldn't we?"

Henke doesn't have a clue about what's going on. Neither do the other officers of *U-515*, who are all being kept isolated in the brig, and they'll stay that way until they reach Norfolk. Gallery makes a prediction: He's not so sure about Henke abiding by the agreement, but those forty or so other petty officers and seamen from *U-515* will soon be singing state secrets.

And for Task Group 22.12, there's one more triumph to come on Easter Sunday.

At the crack of dawn, three turkeys from the *Guadalcanal*—two Grumman Wildcats and a Grumman Avenger torpedo bomber—make radar contact with a U-boat northwest of the Portuguese island of Madeira. The vessel—*U-68*—is riding the surface, apparently in the final stage of a recharge before daylight sends her scurrying underwater.

The planes zero in and launch an all-out attack with .50-caliber bullets and drop their entire loads of depth charges. Then they circle with a stream of rockets blazing and fire some more. By the third run, the U-boat is obliterated. Just like that, she's gone. The planes mark the spot with flares and circle overhead to wait it out. When the sun makes an appearance, the flyers see the unmistakable signs of a confirmed U-boat kill. There, floating on a pool of oil, are three severed human heads. The

Yanks drop a rubber raft and head back to *Guadalcanal* as Dan Gallery dispatches the task group at full speed to search for survivors.

Three hours later, the *Chatelain* spots a dot in the ocean—two German seamen, clinging to a rubber raft. One German is dead and the other is badly injured. Out of *U-68*'s crew of fifty-five sailors, Hans Kastrup is the sole survivor. He's hanging on to the raft with one hand and holding up his lifeless shipmate with the other.

Kastrup and the dead German are hauled on board the *Chatelain*. Kastrup is taken to sick bay, and it's touch and go whether he'll make it. In the meantime, the ship's sailmaker sews the dead German into a canvas sack, with a shell, as a gesture of respect. The body is lowered into the sea for a fitting burial.

In a few days, Kastrup recovers enough to stand on his feet. Captain Gallery sends for him, and Kastrup is transferred by whaleboat to the *Guadalcanal* and brought to Gallery's cabin. With Gallery is the ship's doctor, Dr. Henry Morat, who speaks fluent German and acts as official interpreter.

Gallery eyeballs the lone survivor of *U-68*. He's just a skinny kid, young enough to be Gallery's son. Gallery offers Kastrup a seat.

With Dr. Morat translating, Kastrup proceeds to give a harrowing account of what happened. He says that he and a gunner were on the top deck of *U-68* serving as lookouts when the American planes hit them with everything they had. The gunner was shot in the leg and stomach in the opening seconds of the attack. As the siren sounded a crash dive, Kastrup was struggling to lug his wounded comrade into the U-boat when the conning tower hatch slammed shut before he could reach it. The U-boat submerged and Kastrup and the gunner found themselves pulled under by suction. The next moment, a depth charge blew *U-68* in half. Somehow, Kastrup escaped death. Then a miracle—Kastrup saw the rubber raft that had been tossed out by the Yanks and swam for it with one arm, hauling the unconscious gunner with the other. Kastrup was clinging on for three hours before he was rescued by the *Chatelain*.

Hearing Kastrup's story makes a deep impact with Gallery. Impressive kid, Gallery decides, even if he is fighting for a morally bankrupt regime.

But Gallery points out that when the "going got tough," Kastrup found himself deserted by his skipper. It was he—Daniel V. Gallery of the United States Navy—who stuck his neck out, sent his ships out in submarine-infested waters, and came to Kastrup's rescue.

Kastrup listens "respectfully." Then he looks Gallery in the eye and says, *Kapitän, ich bin deutscher soldaten*—"Captain, I am a German soldier." He explains that in the Kriegsmarine, it is understood that a U-boat skipper is expected to crash dive even if it means abandoning a lookout like Kastrup.

"It was the captain's duty to sacrifice me and save his ship," Kastrup says.

Gallery orders Hans Kastrup back to sick bay. This second-class-nobody-seaman seems to be made of sterner stuff than Werner Henke, holder of the Knight's Cross. (After the war, Kastrup was repatriated to Germany. He and Gallery kept in touch via postcards every Easter.)

■ ■ ■

It's an Easter for the record books. Two U-boats blown up and all those German POWs taken into custody, including an ace of the Kriegsmarine who is apparently willing to turn on the Nazis. Plus, the inauguration of round-the-clock air operations to pound the U-boat fleet into submission.

Gallery sends Atlantic Fleet Admiral Ingersoll a dispatch regarding *U-515*.

"Well done," Ingersoll radios back.

Then Gallery sends a second report, updating him about *U-68*.

"Exceptionally well done!" Ingersoll responds.

Gallery tells the ship's painter to stencil two more swastikas on the side of the bridge.

The USS *Guadalcanal* anchors at Norfolk on April 26. Sometime before midnight, Henke and the crew of *U-515* are ordered to put on blindfolds. Then they file down the gangplank and are turned over to Navy authorities for processing and interrogation.

Now it's time for Captain Gallery to ponder the big takeaways from the air attack on *U-515*. Even a tough son of a bitch like Henke knew when to give up. Henke may be a hero of the Third Reich, but apparently even he is not willing to die for der Führer. When his boat came under siege, he surfaced and jumped overboard. He gave himself and his crew a fighting chance to live out the war.

And how about that German sailor who yelled "Happy Easter!" to his captors? Better to be a live prisoner than a dead war hero.

It gets Gallery thinking about that night long ago, back in Iceland, in 1942. He was in the officers' club with those young American warriors, spinning stories and killing time, when one of the officers threw out a crazy "what-if" scenario: "Why can't we board and capture a sub?"

Gallery remembers walking out and staring at the aurora borealis.

If you could figure out a way of boarding a U-boat before the Germans scuttled her and destroyed the code books and the Enigma cypher machine . . .

Imagine . . . capturing a German submarine . . . with all her secrets intact.

Was it even possible? Back in Iceland, it seemed like a fantasy. But two years later, heading back to Norfolk after blowing up *U-515*, it doesn't seem crazy at all.

EIGHTEEN

SERIOUSLY, DAN?

Preparations start right away for Gallery's next war patrol, Task Group 22.3.

The Pentagon—still not Gallery's favorite place in the world—is where he pays his respects to Fleet Admiral King, and then calls on an old Navy buddy, Captain Henri Smith-Hutton, now assigned to Admiral King's staff as assistant for combat intelligence. He and Gallery have an easy rapport. The two captains have known each other since they were junior turret officers in 1925 on the battleship USS *Idaho*. That's how far back they go. After a few pleasantries, Gallery informs Smith-Hutton that he's about to embark on another sweep of the Azores with a new hunter-killer task group. Then he cuts to the chase: It is his intention to capture a German submarine.

"I've made up my mind," he tells Smith-Hutton.

Gallery gives Smith-Hutton the spiel, based on lessons learned from his encounter with *U-515*. Once a submarine comes under attack and surfaces, he tells Smith-Hutton, there's really no need to pulverize it into pieces straightaway. Make an assumption, goes Gallery's argument, that the German captain and his crew want to ride out the war as POWs. Now the what-if part . . . What if our boys were able to seize the U-boat before the Nazis "pull the plug" and scuttle the vessel to keep her from falling into enemy hands? Imagine the gold mine of intelligence that could be gleaned. The Nazi code books! An Enigma machine! German submarine technology!

As a student of naval history, Smith-Hutton is intrigued. Sounds like a harebrained idea, but then again, maybe not. The last time the American Navy seized an enemy warship in battle on the high seas was in 1815, when the sloop-of-war USS *Peacock* opened fire and boarded the British brigantine HMS *Nautilus* in the Indian Ocean. But that one goes into the record books with an asterisk because the commander of the *Peacock* was unaware that the War of 1812 had ended with the signing of a peace treaty six months earlier.

Smith-Hutton doesn't want to dampen Gallery's enthusiasm, but his friend needs to consider a few troubling truths. He brings Gallery to the Office of Naval Intelligence's Technical Section, where he is shown captured U-boat blueprints that pinpoint the location of the vessel's scuttling valve—a doomsday mechanism capable of flooding the submarine with tons of seawater and sinking her to the bottom of the ocean. All U-boats are also believed to be rigged with fourteen booby traps, scattered in hidden locations throughout the vessel, and set to detonate fifteen minutes after the Germans have abandoned ship.

It sounds daunting. Booby traps. Scuttling valves. Anything else? Perhaps, at this point, Smith-Hutton doesn't know whether he was supplying this bold and unorthodox naval officer with practical information, or just humoring him. Smith-Hutton promises Gallery he'll do whatever he can to assist.

But seize a U-boat on the open sea? Seriously, Dan?

Smith-Hutton has more to show his old buddy. He and Gallery head over to the Main Navy and Munitions Buildings, just off the National Mall on Constitution Avenue, where the Navy was once headquartered before the Pentagon was built. They proceed to the third floor in the Seventh Wing, headquarters for F-21, the Navy's anti-submarine intelligence command, led by Commander Kenneth Knowles.

Gaining admittance to F-21 requires the highest-level security clearance. Very few Navy officers are permitted access to the secret room.

They enter an immense space with oversized wall maps of the Atlantic Ocean. Each pin on the map represents the current tracking position of all known operational Nazi submarines. Sitting at the array of desks is a squad of WAVES—the women's branch of the US Naval Reserve, created by Act of Congress in 1942.

Knowles is outranked by Gallery and Smith-Hutton, but the two senior officers show him the utmost deference. This gentleman is a major player. It was Knowles and his WAVES who steered the *Guadalcanal* pilots to *U-515*'s position on Easter Sunday. And to *U-68*. The information was extraordinarily helpful in narrowing the search zone. A grateful Gallery has taken to calling Knowles "the Soothsayer." Whatever his mix of science, intelligence, art, and even guesswork, Knowles's track record is damn impressive. The Soothsayer has been uncanny in his predictions.

Knowles was born in Nebraska and graduated from the Naval Academy in 1927. He served as gunnery officer on the destroyer USS *Paul Jones* but found himself more suited to academia than life on the high seas. Knowles taught English at the Academy, and in 1936 married Velma Sealy, the daughter of a country doctor from Santa Anna, Texas. When war was declared, Knowles was an assistant English professor at the University of Texas. Then duty called. Knowles reenlisted and was assigned to the Tenth Fleet.

The Tenth Fleet? Every school kid in America during World War II knows the US Navy consists of seven fleets—designated One through Seven.

But the *Tenth?*

In naming it, Fleet Admiral King has skipped numbers eight and nine and assigned himself to its command. The existence of the Tenth Fleet is classified, and it is certainly the most unusual fleet in Navy history in that it has zero ships. Not one cruiser, destroyer, battleship, or aircraft carrier. Not even a PT boat. Not so much as a rowboat. The sole purpose of the Tenth Fleet is to bring intelligence and operations under one command in the hunt for Nazi U-boats. Keep it nimble. Reduce red tape and leaks.

The entire staff of the Tenth Fleet consists of five hundred men and women. That is everyone. No wonder it is also known as the Phantom Fleet. Most of the personnel are WAVES, handpicked for their intellect and flare for cryptography. The men have been recruited from the sciences, academia, the regular Navy and Reserves. They are eggheads, like Kenneth Knowles.

Gallery is in the heart of the heart of it. Inside F-21's U-boat tracking room, the dreamer tells the academic about his audacious scheme to capture a U-boat.

Knowles listens. He is encouraging. Definitely an intriguing notion, he says.

Knowles shows Gallery the wall map. Here's something interesting. F-21 has been tracking an unidentified U-boat since March 16, around the time the submarine had cast off from the French port of Brest. Knowles and his staff had not paid much attention to the vessel until March 25, when F-21 intercepted a radio transmission from the U-boat skipper updating Doenitz's headquarters on his current position.

The submarine is southbound, on her way to hunt prey off the coast of Africa.

Doing the math, Knowles figures the U-boat will be at sea for three months before she starts running low on fuel and heading back to France. March. April. May. That might time out perfectly for Gallery, because Task Group 22.3 will be in the same general vicinity of the Atlantic by the end of May.

Gallery likes what he's hearing. Could that U-boat be just what he is looking for?

They bid one another farewell and good hunting. Gallery proceeds on to Norfolk and the USS *Guadalcanal*, leaving Commander Kenneth Knowles with a lot to think about.

Capturing a German U-boat. The audacity! Knowles recognizes the intelligence windfall of such an accomplishment. Obtaining the current Kriegsmarine code books, which for security purposes change every two weeks, and the latest model of the Enigma cipher machine. Knowles sure would like to get his hands on one of those new acoustical torpedoes the Germans have started deploying. They're wreaking havoc in the Atlantic. There might be some way of neutralizing the weapon if the Allies could figure out what made it tick.

Of course, Knowles couldn't let Dan Gallery in on everything, such as the fact that a brilliant mathematician from England named Alan Turing and his team of cryptographers at Bletchley Park had cracked Nazi Germany's military code. Churchill knew. Also Roosevelt. And Eisenhower. A small number of other top military brass.

And Kenneth Knowles.

It is the *Ultra* operation. Strictly on a need-to-know basis. And Dan Gallery didn't need to know.

NINETEEN

RADIO SILENT

U -505 departs Brest, France, at 6:35 p.m., under cover of darkness and mired in a somber mood. The crew can't help wondering if their time is running out. They figure they've got maybe a one-in-three shot of making it back to France.

Kapitänleutnant Harald Lange opens his sealed orders and announces that the submarine will be setting off for the coast of West Africa. A navigational chart is laid out, and the men eagerly surround it to see where they are heading.

It's going to be a long haul.

As the weather turns warm, the sailors do what they can to bide time. Hans Goebeler reads Jack London stories, in English. Alaska fascinates him, made vivid by London's stories about survival in the bone-numbing cold, so different from the stinking heat of the submarine.

At night, as *U-505* picks up speed running the surface with her diesels, Lange permits the men to enjoy a smoke and gaze at the stars above them. Always, the watch officers on the bridge keep their eagle eyes trained on the sky for enemy planes. It's a nice break when Toni the cook brings them a pot of hot coffee mixed with rum.

On March 25, 1944, two weeks after casting off from Brest, Lange radios his position to U-boat headquarters in Berlin. Lange, and even the German High Command, are oblivious to the fact that the Kriegsmarine's military code has been cracked by Allied military intelligence.

About 3,600 miles away, *U-505*'s location is duly noted on the tracking map by US Navy Commander Kenneth Knowles's squad of WAVES at the Old Navy and Munitions Building in Washington. Based on the US Navy's decryption of *U-505*'s last radio transmission, the boat is at this moment positioned at 44-39N 14-30W.

Then, for the next five weeks, *U-505* goes radio silent.

TWENTY

THE NINE

I n Norfolk, Virginia, Captain Gallery is about to set sail with the newly formed Task Group 22.3—and he's got a bombshell to drop on his men before they even leave the harbor.

The skippers of five destroyer escorts that are to accompany the USS *Guadalcanal* listen attentively as Gallery and his XO, Jessie Johnson, conduct a predeparture confab in the aircraft carrier's ready room.

Gallery lets Johnson do most of the talking.

Then Gallery takes the floor.

Here goes.

Gallery informs his commanders that he's been thinking for a long time about seizing a U-boat and towing her back to America.

Ignoring the slack jaws and bug-eyed stares, he proceeds to outline his plan of attack.

"When and if we bring a sub to bay, we don't clobber it forthwith as we had with *U-515*. Instead, assume the German skipper surfaces for the sole purpose of saving the hides of his crew and intends to scuttle as soon as the crew gets overboard. When he surfaces, we will therefore cease fire with ammunition that could do serious structural damage to the boat."

In other words, no depth charges, torpedoes, or rocket fire. Nothing that will pulverize the boat and blow her out of the water.

Instead, Gallery continues, "We will blast away briskly with small-caliber antipersonnel stuff in order to keep them away from their guns and encourage them to get the hell off that U-boat so we can put an inspecting party on board."

The "inspecting party" will consist of US sailors who will seize the boat, climb down the submarine's hatch, close the scuttling valves, disarm all booby traps, and do "whatever else they have to do" to keep the U-boat afloat.

"We will then pass them a towline and bring the sub back to the United States."

Gallery is done. He lets everyone absorb all that he's said.

Total dead silence. As Gallery will later recall, he wonders if his officers think he's got "barnacles on the brain." It is so quiet, a "flake of falling dandruff would have made an audible noise."

There, in the ready room, sit the commanders of his five destroyer escorts—the *Pillsbury*, *Chatelain*, *Flaherty*, *Pope*, and a newcomer to the group, the USS *Jenks*, named in honor of a young lieutenant junior grade, Henry Pease Jenks, who was killed when his cruiser was torpedoed during the naval Battle of Guadalcanal in 1942. *Jenks*'s skipper is a young commander, Frederick Hall, from Ohio.

Looking out at the five commanders and their XOs, Gallery sees them exchanging skeptical glances at one another. Capture it *and* tow it home like King Kong? In the back of the room, he observes an officer pointing at his head and making circular motions with his extended finger, as if to say, *Have you ever heard anything so nutty in your life?*

Gallery lets it roll off his back, and he has one more thing to add.

"I want each ship to organize a boarding party and keep a whaleboat ready to lower throughout the next cruise. Also, keep your towline where you can get to it in case we need it."

His pitch is done. He smiles at the men. "Any questions?"

A couple of the skippers are about to speak, but then reconsider. Nobody wants to question the good sense of their commanding officer during a predeparture briefing.

"There being no questions, the meeting is adjourned."

The five commanders and their XOs return to their ships.

Mother's Day, May 14, 1944, brings fresh hope from anxious parents across the United States that the war could be in its final year. The invasion of Western Europe is close at hand. When and where, no one can say. The next day, Monday, May 15, 1944, Clyde Shoun of the Cincinnati Reds pitches a no-hitter against the Boston Braves. There's not as much fanfare as there'd usually be for such an accomplishment, as most of baseball's best players have been called to serve Uncle Sam. Far away in Budapest, the pro-fascist government of Hungary, under the direction of the SS, launches the systematic deportation of 440,000 Jews from the nation. Most will be sent to the death camp at Auschwitz. The annihilation of the Jewish population of Europe continues apace.

And on this day, Task Group 22.3 clears the Virginia Capes and sets sail for the Cape Verdes Islands off the coast of West Africa to hunt and kill Nazi submarines.

Gallery sends a signal to his five destroyer escorts, reminding each commander to initiate the task of organizing their respective boarding parties.

WILL BE HOT ON TRAIL TOMORROW X EACH ESCORT DRAW UP PLANS AND ORGANIZE A PARTY TO BOARD CAPTURE AND TAKE SUB IN TOW IF OPPORTUNITY ARISES.

■ ■ ■

On the USS *Pillsbury*, Lieutenant Commander George W. Casselman has a top candidate in mind to lead his boarding party—the ship's chief engineering officer, Lieutenant Junior Grade F. M. Burdette. But then Casselman reconsiders. He calls a conference of his officers, fills them in on Gallery's objective, and asks Lieutenant Junior Grade Albert Leroy David if he'd like the job.

The bespectacled David is an interesting choice. He is a "Mustang," military slang for a commissioned officer who started his career as an enlisted man. (The Brits call them "Temporary Gentleman.") Despite his junior grade rank, David is no kid. He was born in 1902, making him forty-two. He's a landlubber by birth, raised in Maryville, Missouri. He enlisted in the Navy in 1919 when he was seventeen, reenlisted in 1921, and served on a succession of ships, including the USS *New York*, *Texas*, and

Trenton. He reenlisted in 1931 and then transferred to the Navy Reserves when his twenty years were up. In a nutshell, a career Navy enlisted man. In 1939, after Hitler invaded Poland, he was recalled to active duty.

David was a confirmed bachelor until he fell in love with a forty-year-old practical nurse named Lynda Kleber. They met in 1942 at the University of Wisconsin, where the Navy had sent David to study diesel engines. It was a whirlwind romance, as happens in time of war. He and his pretty brunette drove to Dubuque, Iowa, and got married before a justice of the peace in December. For the middle-aged lieutenant who had spent his life at sea, it was quite an adjustment because Lynda had a young son and daughter from her first marriage to an auto mechanic.

Lieutenant Commander Casselman is impressed with David's long years and breadth of Navy training. He knows hard knocks, sticky situations, and how to speak the lingo of enlisted personnel because up until 1942 he'd been one. Most of all, David knows a lot about everything: engines, diesels, machinery, electricity. He is even acquainted with submarines, having served five months' duty as a machinist in the Navy's Submarine Repair Unit in San Diego. That submarine experience, Casselman decides, is what makes him the superior candidate over Burdette. Burdette's name is crossed off the boarding party roster and replaced by Albert David.

David starts making the rounds on the *Pillsbury* recruiting volunteers. Zenon "Zeke" Lukosius is at his station in the engine room when the lieutenant comes around and makes an announcement: "We're going to capture the next sub."

If there's one thing Lukosius has learned in the US Navy, it's never volunteer for anything, no matter what. All the same, the kid from Chicago steps forward.

"I'll volunteer."

David takes Lukosius's name down and says he'll be in touch. Then he goes looking for other guinea pigs.

As David makes his exit, a sailor standing next to Lukosius has a question. Did Lieutenant David just say seize a Nazi submarine on the high seas? Sounds like total lunacy. "Why'd you volunteer for something like that?" the sailor asks.

"I'll tell you why," Lukosius answers. The last time he was in a battle, it was the day *U-515* fired that acoustical torpedo at the *Pillsbury*. Lukosius was stuck in the engine room listening to the excitement over his headphones. He didn't like being benched during the big game. Epic events are taking place in the world, and Lukosius wants to take part in the action.

"I don't want to be in the engine room," he declares. "I want to be up there either shootin' at 'em or catchin' 'em."

There is another factor, which Lukosius has to admit with a chuckle. Maybe, just maybe, he has "no brains."

■ ■ ■

Signalman Second Class Gordon Hohne figures he has the best job in the Navy. He's on the bridge pretty much from sunrise to sunset in his whites because that's the way the Navy likes its signalmen to dress, so they stand out. He's thinking about making the Navy his career.

Hohne reads a sign-up sheet recruiting sailors for the *Pillsbury* boarding party and figures, *what the heck, I'm single*. He puts his name on the dotted line.

One by one, the boarding party candidates are winnowed down to nine sailors. Each man is selected for a unique skill set he brings to the mission's objectives.

They are:

Lieutenant (jg) Albert David, boarding party commander.

Gordon Hohne, signalman second class. He is to position himself in the U-boat's conning tower and maintain continuous communication with the USS *Guadalcanal* and the *Pillsbury* via semaphore flags.

George Jacobson, chief motor machinist's mate. An experienced ship's engineer, he is to secure the submarine's main control room and stabilize all critical values and controls.

Arthur Knispel, torpedoman third class. The muscle. He is to carry a Tommy gun and engage any German sailor who resists the order to surrender. Knispel is designated to be the first American to climb down the U-boat hatch and penetrate the submarine. If there are any Germans present, and they open fire, Knispel will almost certainly be the first man cut down.

Zenon Lukosius, motor machinist's mate first class. Secure the control room and, additionally, secure the "sea strainer" water valve, which is designed to flood and scuttle the submarine before the Yanks seize full control.

Chester Mocarski, gunner's mate first class. Supplemental muscle. If necessary, he is to use lethal force and eliminate any German sailor found inside the submarine.

Wayne Pickels, boatswain's mate second class. Jack-of-all trades. Supervise the boarding of the small whaleboat that will ferry the boarding party to the Nazi submarine. Once inside the sub, he is to secure the Enigma machine and locate any code books left behind by the Germans.

William Riendeau, electrician's mate third class. Secure the U-boat's electrical power control station and assist in locating and disarming fourteen booby traps connected to the U-boat's internal circuitry.

Stanley Wdowiak, radioman third class. Seize the Enigma machine and related communications material. Secure radio shack. Assist in neutralizing booby traps.

They come from all walks of life and backgrounds, but with a common thread—the courage to just go for it.

Gordon Hohne is the "baby" of the group, and he's not only defying the odds—he's defying his mother's wishes. He grew up in Worcester, Massachusetts, in a neighborhood teeming with Swiss and German immigrants. His middle name is "Fritz" and his parents, who came from

Germany, work at the Norton Company manufacturing grinding wheels. Hohne made his first stab at joining the Navy in April 1942 at age sixteen, but his mother, Frieda, refused to sign the enlistment papers because, from her point of view, her son would be fighting against her homeland. It strained their relationship. When he turned seventeen, Gordon tried again, and this time he got his father to consent. Hohne was sent to boot camp in Newport, Rhode Island, for three weeks where he slept on hammocks. After taking an aptitude test, he was assigned to Signal School on the campus of Butler University in Indiana. He came in seventh out of his class of almost two hundred sailors.

Chester Mocarski is the proverbial kid from Brooklyn. He's a big guy—six feet tall and muscular, at 185 pounds, with a ruddy complexion and blond hair. His parents are Polish immigrants. Mocarski dropped out of high school after two years and was working as a machinist's apprentice when he got a jump start on the war and enlisted in the Navy at the 1st Avenue Armory in Brooklyn. This was three months before Pearl Harbor. When he volunteered for the boarding party, he'd just returned from shore leave visiting the family in Brooklyn—and marrying his girlfriend, Veronica.

Stanley Wdowiak is another Brooklyn boy, shorter than Mocarski at five-foot-eight, but just as tough. His parents are Polish immigrants who came to America in search of a better life. They live in the Greenpoint section where, it is said, the heart of Poland-in-America beats. His father is an iron worker, his mother a midwife. Stanley dropped out of Catholic school to enlist.

George Jacobson is the dinosaur of the boarding party—at age forty-four, he's got two years on Albert David. He's a beefy six-footer from Portland, Oregon, weighing two hundred pounds, with black hair and gray eyes. Before the war, he was a merchant marine with Shell Oil. His father came from Denmark and worked as a shipyard laborer. Jacobson's been around so long he still has his draft card—from World War I.

William Riendeau is nineteen and has a thick New England accent. Back home in Rhode Island, he's got two kid brothers who are itching to sign up.

Arthur Knispel is also nineteen and wiry, lean as an alley cat at five-nine and 155 pounds. He was raised in Newark, New Jersey, where his father is a baker and his mother a bookkeeper. He's got blond hair and blue eyes and is the youngest of three siblings. Arthur never got to high school, wrapping up his schooling with an eighth-grade education. His parents are from Poland, proud as they can be of their son. After all, he's doing his part to right a wrong that began when Hitler invaded their native land.

Many of them go by their Navy nicknames. Wdowiak is "Stash." Hohne is "Fritz," or sometimes "Flags," due to his skill with semaphore flags. Pickels is "Mack." Lukosius is "Zeke."

Lieutenant Commander Casselman looks over the list. A first-rate company of sailors. Maybe they volunteered to score brownie points and don't think this cockamamie undertaking will ever happen, but he had to give them credit for stepping up.

Training begins right away, with Lieutenant David in charge.

David orders the boarding party to meet way back in the fantail, where they can drill and practice climbing into the whaleboat with all the heavy gear and weaponry they'll be taking with them. There's no dawdling. One major objective is cutting down the response time. When the call to action comes, they'll need every second to hustle to the whaleboat, load up, and speed off in pursuit of the Nazi submarine.

There's homework, too. Each member of the boarding party is provided a written set of instructions laying out their duties and assignments. The title: BOARDING AND SALVAGE BILL FOR ENEMY SUBMARINES.

The object of this bill is to have available a group of men aboard who by study and training have so familiarized themselves with the layout and location of the most important equipment, flooding and trimming valves and cocks to facilitate quick and positive maneuvers to forestall the enemy from scuttling his ship.

Even though the opportunity may present itself very rarely, a well-organized crew, each man thoroughly familiar with his specific duties, can make the difference between valuable capture and "the one that got away."

The sailors will know to report to the whaleboat when they hear the call, "Away submarine boarders, away."

All members of the boarding party will be armed with a .45-caliber automatic pistol, ammunition belt, flashlight, and gas mask.

They won't be traveling light, as they'll also be taking six hand grenades, six tear gas grenades, one Tommy gun, and a machinist's toolbox with T-wrenches, four Stilton wrenches, and other utensils. Also, a ten-pound sledgehammer, a crowbar, and a thirty-foot steel chain.

David is designated the boarding officer. Once inside the sub, his orders are to proceed to the wardroom and pry open two safes. They have a pretty good sense of the U-boat layout, based on the capture of *U-570* in 1941 and *U-464* in 1942 (for more information, see chapter 2), and naval intelligence reports from before the war. They know that the most important safe is located above the U-boat captain's bunk. Another safe is below a small built-in desk on the port side of the forward bulkhead.

And what they contain are documents worth their weight in gold.

Getting these publications is very important and haste is necessary as the submarine may remain afloat for only a limited time.

Torpedoman Knispel will serve as armed guard.

He goes down conning tower hatch first. He shall provide and carry Tommy gun and shall be ready and prepared for all types of strong-arm methods. After cleaning out the main control room and wardroom he shall go forward to the torpedo room looking for planted charges, and to secure all torpedoes.

The gunner's mate, Mocarski, oversees all hand grenades and tear gas.

If there is evidence of resistance (he is) to create an intense light inside the submarine to temporarily blind any of the enemy stationed below.

Signalman Gordon Hohne will carry a signal gun and his semaphore flags. His review of the set of instructions isn't inspiring a lot of faith in the mission. For him, it's just a "sheet of paper" and a whole bunch of "mumbo jumbo." Seizing a Nazi submarine on the high seas? As if that's really gonna happen.

As Zeke Lukosius looks over the document and studies his assignment, he finds it baffling.

He shall go below and immediately close the main vent valve. Number one located on the port side of the after bulkhead of the control room.

Lukosius shakes his head. *Main vent valve?* He has no idea where that is. He reads further.

He now closes vent valves #2, #3, #4. These valves are levers located in the overheard in the after part of the control room. To close, push them up and secure them.

It's getting really complicated. He wants to ask Lieutenant David, "Now how in the world am I gonna find control valves two, three, and four?" But he holds his tongue, figuring he'll either be fully prepared when it's game time, or wing it. He knows that he must be prepared for surprises galore.

It's practice and more practice. Sometimes they skip a day. Sometimes a few days go by before they assemble for another drill. They also shorten the battle cry to something simple: "Away Boarders!"

To their pals on deck who are watching all this fuss play out, it's getting to be a source of amusement. What are these guys up to anyway? It all seems like a bunch of baloney.

One day after an exercise drill, Stanley "Stash" Wdowiak runs into Lieutenant David and asks him a no-bullshit question with classic Brooklyn bluntness:

"Do you think the Germans are just going to let us come aboard? Huh?"

David dryly responds, "Well, accidents will happen, Stan."

When he has a moment to relax in his bunk, Pickels takes out his diary. He hasn't written anything in a while. Too damn busy. One feature of the boarding party operation is really stressing him out. It sounds daunting, and, wouldn't you know it, that's his assigned task. He must lug the thirty-foot-long, three-eighth-inch chain onto the U-boat. It may weigh more than he does.

On boarding (he) shall lash one end of the chain to the deck, dropping end with toolbox attached down to conning tower hatch, preventing any remaining enemy from closing hatch and trapping boarding party below.

Pickels has to laugh. Yeah, right. Easy enough to write it up when nobody's shooting at you.

Just thinking about it is so unnerving . . . trapped inside the enemy submarine, with no way out, no chance of escape. Surrounded by Germans bent on slaughtering them.

Albert David
Gordon Hohne
George Jacobson
Arthur Knispel
Zenon Lukosius
Chester Mocarski
Wayne Pickels Jr.
William Riendeau
Stanley Wdowiak

Six out of nine are sons of immigrants; all are Americans to the core. They have a lot to think about. No one really knows what to expect. Are they about to embark on a suicide mission?

Could be.

TWENTY-ONE

JUST ONE MORE NIGHT

To hell with those acoustical torpedoes! Hans Goebeler has had enough of the finicky wonder weapons.

Every torpedo tank on *U-505* is now fully armed with the state-of-the-art eels, which can sink an enemy destroyer even if poorly aimed. All a U-boat skipper has to do is fire in the general direction of the enemy ship and the torpedo will home in on the sound of the propellers for a bull's-eye. At least, that's the way it's supposed to work, but the reality is a little different. These torpedoes are temperamental beasts and require constant maintenance. They must be pulled out of the tubes every day and mopped dry, and the fuse mechanism bled and adjusted; otherwise the buildup of moisture could corrode the electrical firing fuse, or even lead to an accidental detonation and blow up the submarine. What a farcical way to die—obliterated by your own torpedo.

It is April 27, 1944, and *U-505* is just two nautical miles off the coast of Liberia, the African nation founded in 1822 by freed American slaves. Goebeler is indifferent to Liberia's special place in history. All he cares about is that Liberia is the main source of natural rubber latex for the Allies and a key destination for American and British merchant ships.

For a skipper as risk averse as Kapitänleutnant Harald Lange, navigating this close to the port of a foreign adversary is a daring maneuver. The U-boat hull clears the water depth with only a few meters to spare, and there is danger of running aground. But the Old Man knows these waters well from his days in the merchant marines. From this distance, Lange can make out the lights in the deep-sea harbor of Monrovia, the capital city named after the fifth US president, James Monroe. Lange counts twenty or so piddling fishing vessels of limited military value. Not worth a torpedo, or even a single potshot from the mounted antiaircraft .105mm cannon. Time to move on.

It has been a frustrating month. Ever since departing Brest and successfully dodging enemy patrols as they crossed the Suicide Stretch, a.k.a. the Bay of Biscay, then entering the patrol zone off the coast of West Africa, *U-505* has not encountered a single enemy ship. It's as if the Allies know of their presence and warned convoys to steer clear.

With Monrovia turning into a bust, Lange navigates *U-505* to Cape Palmas on the extreme southeast corner of Liberia, a sea lane ripe with enemy ships making their way past the Horn of Africa toward Great Britain.

Nothing.

It's empty out there.

At last, on May 10, a British freighter is spotted off the stern, and she's a big one—nine thousand tons. A "fat morsel" of a ship, Lange writes in his war diary. The skipper wheels *U-505* around and gives chase on the surface at flank speed. The freighter is a fast ship, but *U-505* hasn't gone more than four thousand miles to turn back empty-handed. The submarine is gaining on her, with the diesel engines pumping away at a two-knot advantage and bound to catch up. Then Lange spots a destroyer—probably British—way off on the horizon, but closing in, following *U-505*'s exhaust fumes.

From hunter to hunted, in the blink of an eye.

Lange abandons the chase for that juicy freighter and dives to periscope depth. Meanwhile, the commander of that British destroyer won't steam any closer than within five miles of *U-505*, which keeps it far out of torpedo range. Lange shakes his head. Clever bastard, that Brit. Next time.

Lange retreats and resumes sailing along the African coast. Again, nothing. Not a single torpedo fired during the entire war patrol! By the end of May, Lange has had enough. He orders *U-505* to turn north. It's time to head back to Lorient. Maybe they'll get lucky and run into Allied ships along the way.

"What a dismal trip," machinist Hans Decker bemoans.

■ ■ ■

About 350 miles from *U-505*'s current position, Captain Daniel Gallery, on the USS *Guadalcanal*, also finds himself weighed down with exasperation.

For three weeks he has been on the hunt for German U-boats, but so far, no submarine has come within his sights, and he is running out of time.

He's scouring the Atlantic Ocean around Cape Verde Islands, prime hunting grounds for U-boats on their way to and from the Bay of Biscay and the African ports of Freetown and Monrovia. He's where he's supposed to be, but all he runs across are whales, pods of porpoises, Portuguese fishing trawlers, and, most maddening of all, a whole bunch of false alarms and those mythical creatures known as "electronic gremlins" who catch the blame for problems with no logical explanations.

Three thousand yards ahead of the *Guadalcanal* is his protective screen of five destroyer escorts consisting of the *Pillsbury*, *Chatelain*, *Flaherty*, *Pope*, and that newcomer to the task group, the USS *Jenks*.

In the air, scouring the ocean for 160 miles on each side of Task Group 22.3, are four turkeys on round-the-clock patrol. By now, every pilot on the *Guadalcanal* is certified in night flying. Gallery has made sure of that. They're as confident as owls about landing in total darkness on a pitching postage stamp in the ocean.

Gallery is in the Combat Information Center, situated on the main deck of the *Guadalcanal*. There are no windows in CIC, but you can see more here than on the brightest day on the bridge. This is the brains of the ship, jammed with radar, maps, scopes, radio transmitters, and a squawk box to communicate with his pilots. It reminds Gallery of a broadcast studio, or an illegal bookmaking establishment. Maybe the closest comparison is to an air traffic control tower. A large vertical plotting board displays the current position of all American and British planes in the air as well as all ships, be they friendly or hostile. The TBS box (Talk Between Ships) allows Gallery to communicate with the commanders of his destroyers via secure transmissions. He keeps the gabbing down to a minimum, limiting the conversation to urgent tactical commands. A simple "Five by Five" response will do, which means, in Gallery-speak, everything is "hunky-dory."

As the planes come in for landing, it's turning into another vexing day for Gallery. Once again the pilots are returning empty-handed, and they're starting to wonder if this mission is a gigantic waste of resources and time. The same patrols, back and forth, all day long, combing the same ocean space and finding nothing.

They're young and impatient. Gallery was once that way, too. He reminds everyone that hunting submarines is 99 percent tedium for that 1 percent payoff. Just because they're not catching sight of the submarines visually or making contact with radar or sonar buoys, that doesn't mean there aren't any submarines out there. It just means the U-boat skippers are growing more skilled at evading detection. They've learned that the deeper you dive into colder water, the more difficult it is for aerial radar to penetrate. The trick is to catch the U-boat when she surfaces, which she must do at some point to recharge. Then, even if she crash dives, you've got that location vectored and you concentrate your search area where the sub was last spotted. A submerged submarine can only advance sixty miles or so before she runs out of juice. It's only a matter of time before she comes to the surface. It's all about tightening the noose. And not giving up.

But time is running out for Task Group 22.3. Gallery's chief engineer and number one "nudge," Commander Earl Trosino, is pestering him. The *Guadalcanal* is running low on fuel. Gallery needs to abandon the dream of catching a sub and set sail for Casablanca to refuel. The way Trosino sees it, making naval history must be put on the back burner when you have to consider banal matters, such as not running out of fuel in the middle of the ocean in time of war.

Trosino can be a pain in the ass, but Gallery has to admit he's the best there is when it comes to engines. He's coming around to Trosino's thinking, to the point of calling it quits.

The date is May 29. As Gallery deliberates, a coded radio dispatch is handed to him.

Damn.

A German U-boat just torpedoed the aircraft carrier USS *Block Island.*

Block Island is *Guadalcanal*'s sister ship. Like *Guadalcanal,* she's a baby flattop, having launched two years ago in Seattle. The sinking happened off the Canary Islands, about 870 miles south of *Guadalcanal*'s current position near Cape Verde Islands. *Block Island* was leading hunter-killer Task Group 21.11 when the German submarine *U-549* somehow slipped past the carrier's destroyer escort screen and opened fire with three torpedoes. Two scored direct hits. A third torpedo fired eight minutes later delivered the coup de grâce. *Block Island* went down in the Atlantic at 9:55 p.m.—the first American aircraft carrier to be sunk in the war. Six sailors were killed and the remaining crew of 951 men were picked up by destroyer escorts. Six Wildcat fighter planes were in the air when the carrier was lost and had no place to land. They changed course to the nearest land base, which was the Canary Islands, but they didn't make it and had to ditch in the Atlantic when they ran out of fuel. Only two of the pilots could be rescued. What a disaster.

U-549 wasn't done yet. She fired another salvo of acoustic torpedoes and hit the destroyer escort *Barr,* leaving her dead in the water. Later that evening, vengeance came when the USS *Ahrens* sank *U-549.* All fifty-seven German hands were lost.

A day of courage and chaos on both sides, and it really hits home with Gallery. *Block Island* was engaged in the identical hunter-killer submarine operation as Task Force 22.3. How on earth does a U-boat slip through an escort screen consisting of four destroyers and sink an aircraft carrier? What the hell happened?

The next morning, Gallery orders his sailors to muster on the *Guadalcanal* flight deck. They have the right to know the good and bad in their theater of war. They'll hear about it anyway, so best to hear it directly from the boss. Plus, it reinforces what's at stake and keeps every-one on their toes. Gallery makes the announcement that *Block Island* is gone. He stares at his assembled crew.

"Does this news scare us?" he asks.

He lets fifteen seconds pass.

"I can see the answer in your faces, and the answer is, 'Hell, no!'"

Gallery dismisses the men before anyone dares say the real answer aloud.

Another urgent coded message is handed to Gallery, this one is from F-21, the Navy's top-secret headquarters for anti-submarine intelligence. Commander Kenneth Knowles, a.k.a. the Soothsayer, is on to something big. There's a U-boat out there all right, says the Soothsayer. It's the same U-boat that Knowles had informed Gallery about all those weeks ago when he was in Washington. F-21 has been tracking the submarine since March 16 when she cast off from the port of Brest, France, on her way to hunt prey off the coast of Africa. Now, the boat has completed her war patrol and is homeward bound to France. As predicted, the strategy is timing out perfectly for Gallery. The U-boat is currently positioned three hundred miles or so north of the USS *Guadalcanal*. Knowles can't be any more precise about the ship's longitude or latitude. As a moving target, the U-boat's position can only be approximated. But she's definitely within striking distance of Task Group 22.3.

This could be what Gallery has been waiting for.

Gallery makes a decision. Today is May 30. If he focuses the entire task group's operations on that U-boat bound for France, they should run smack into each other on June 2.

Of course, there's that dicey business with the fuel supply. Earl Trosino is going to blow a gasket when he hears about this.

For the next three days, the Soothsayer feeds Gallery a continuous stream of tracking intelligence regarding the U-boat's voyage home. At this moment, the vessel—whose identity is not known—is still heading east, straight into the path of the *Guadalcanal*. Gallery orders the com-manders of the five destroyer escorts in Task Group 22.3 to conserve fuel and operate at half capacity so they can wait for the submarine to come to them.

On June 2, planes from the *Guadalcanal* report making radar contact with an unidentified craft. A submarine? Maybe.

Later that day, sonobuoys dropped into the ocean pick up the sounds of a U-boat's propellers.

Then a radar blip appears on a patrol plane's scope, only to promptly disappear, indicating the sub is refueling, and submerging when she realizes she's been made. Whoever that U-boat skipper is, he's a "very cautious" and seasoned professional, Gallery concludes.

On June 3 at 1:00 a.m., American pilots flying overheard detect another "noisy" sonar buoy. They're so certain it's the sub, and since there's no time to waste if it has ill intent, they drop a payload of bombs and torpedoes over her last reported position.

When Gallery wakes up at sunrise on June 3, he half-expects his planes to have sunk the U-boat. No such luck. There's no evidence of a kill—no pool of oil or U-boat debris over the spot where the planes dropped their bombs five hours ago.

Gallery and the senior officers of the *Guadalcanal* crowd into CIC. Analyzing the information, it can only mean the German skipper was driven down before he was able to recharge his batteries.

"He has to come up soon," Gallery says.

He's sitting in the captain's chair on the bridge of *Guadalcanal* when he sees Earl Trosino approaching with a deeply distressed expression on his face. Here it comes. *Jesus*, Gallery must be thinking, *not again.*

Gallery and the wiry five-foot-seven Italian American from Pennsylvania may be about the same size but they are not seeing eye to eye. They've been butting heads since *Guadalcanal* set sail a year ago in Oregon. Gallery knows that his chief engineer has had enough of the "Fighting Irishman" and wants off the *Guadalcanal*. With three stripes on his commander's uniform, Trosino is an old salt who has more seniority than any other chief engineer in the Navy's fleet of baby flattops.

He digs in his heels and gives it to Gallery straight.

"Cap'n, we've got to quit fooling around here and get to Casablanca. I'm getting near down the safe limits of my fuel."

For the next few hours, Gallery and Trosino are at each other's throats. It reminds Gallery of the back-and-forth they exchanged not long ago over the top speed *Guadalcanal*'s engines could achieve. Trosino pegs it at 200 rpm. But Gallery knows a seasoned chief engineer like Trosino always keeps at least ten, maybe fifteen revolutions up his sleeve. Same with fuel. In his gut Gallery suspects Trosino is being cagey and has a "little bit extra fuel stashed away for emergencies."

Trosino isn't budging. Plus, he's insulted by the insinuation. "Cap'n, I don't have any fuel," he insists. He says he would never deceive the commander of a ship about fuel because "if they ever found out, there'd be my neck." Trosino tells Gallery he always gives a "true reading."

So, Gallery shares his own true reading with Trosino: If we can stay on the hunt just a few more hours, that U-boat skipper's goose is cooked.

Trosino blinks. "Sir, may I suggest something?" *Maybe*, the chief engineer concedes, there's enough fuel for one more day of operations if Gallery reduces the task force speed.

"I can give you twenty-four more hours of operations, but beyond that you're finished. *We're* finished." He also wants Gallery to understand there's a good chance he will pull into Casablanca with the tanks bone dry. Hell, he may even exhaust his fuel in the ocean and have to radio for a tug to tow him in, which, Gallery acknowledges, would earn him a place in Navy history he'd never live down.

Gallery accepts Trosino's recommendation. "I want another crack at that bastard," he says.

Trosino returns to the engine room. It's just below the waterline, on the same deck as the mess hall, barbershop, shoe repair, laundry room, and sleeping compartments. There's even a soda fountain and an emporium where they sell ladies earrings, brushes, and watches for when the sailors go on shore leave. Top side is for the glamor boys, the aviators. Below deck is where the unsung heroes of the Navy sweat it out in the engine room and keep the ship running. Trosino is back in his element now, carrying a monkey wrench tucked into his pocket and a rag to wipe the grease off his hands. Maybe he's upset with himself because he knows Gallery has browbeaten him into conceding there's enough fuel for one more day of operation. It's not the first time he cusses that Fighting Irishman.

■ ■ ■

At sunset, on the bridge, Captain Daniel Gallery orders the task group to turn south and double back over the span of ocean they patrolled the previous night. The sky is filled with planes from the *Guadalcanal* on the hunt for that son-of-a-bitch U-boat.

Just one more night is all he needs. That Nazi skipper is in a lousy situation, and he has to surface soon or he's done for. Assuming Gallery doesn't run out of fuel first.

TWENTY-TWO

BATTLE STATIONS

Kapitänleutnant Harald Lange might not make it back to France. He's low on fuel, but the real source of his anxiety is the steady corruption of the boat's storage batteries.

"They're in bad shape," says his machinist, Hans Decker.

Life on board *U-505* is becoming intolerable. A dense and clammy fog is engulfing the vessel due to heat from the equatorial waters, and the steamy air the crew breathes is turning noxious. It's almost like inhaling smoke.

In the sky, enemy aircraft are a relentless presence, scanning the ocean for submarines. Where are these fighter planes coming from? Lange has a sickening feeling that it can only mean one thing: An aircraft carrier is in the vicinity. He's gripped by a soul-sinking realization that a hunter-killer task force is on his tail—and they are zeroing in.

On June 3, Lange makes his move. It's a daring ploy, and it comes about because he's in such a tight spot. Lange orders the water tanks blown, and *U-505* rises to the surface. When he climbs the conning tower ladder and opens the hatch, the air has never tasted sweeter and the sky is gloriously bright—no clouds, and by some miracle, no planes. Now, if he can only run the surface for the next few hours and keep the electric motors spinning so that he can bring about a full recharge of his depleted batteries.

The diesels start pumping away and fresh air sweeps throughout the boat as *U-505* sprints across the Atlantic. During the next few hours, they see Allied planes high above them but somehow *U-505* escapes detection. Lange's lookouts are at the ready, manning the U-boat's antiaircraft armaments, although as any gunnery officer will tell you, hitting an enemy plane from a swaying submarine deck is no easy feat. The weaponry is also of limited value because cannons positioned on the top deck corrode due to perpetual exposure to salt water.

Lange orders a course due north, hugging the west coast of Africa. There is no question in his mind that the hunter-killer task group on his tail must be operating farther east. Of course, he has no idea that the bearing he is charting is taking him straight to Captain Gallery and the *Guadalcanal*. As Lange turns north, Gallery turns south. They are in a direct intercept course.

To conserve oxygen the next morning, Sunday, June 4, every crewman not on duty is ordered confined to his bunk. What a miserable way to start the day. Hans Goebeler lies there waiting for time to pass before his shift starts. By noon, he is back in the control room.

Suddenly, they hear a raking noise, as if a steel cable is scraping across the top deck. What is that thing? It stops, then resumes again. Then the soundman hears the truly forbidding noise of a propeller, off the stern. And not just one propeller. *Three* sets of propellers.

Lange has just finished lunch in his cabin and is taking a nap when he's awakened and told they have a serious crisis on their hands. He hustles to the control room—still wearing black silk pajamas—and makes a quick assessment.

"Battle stations!" Lange calls out. "Torpedo!"

As the men scramble to their battle stations, Lange proceeds to periscope depth to investigate, but with extreme caution. He doesn't want his movement churning up the ocean.

Enemy sonar will detect any turbulence.

"Make it slow, Chief. Maybe we're in the middle of a convoy."

He fully intends to let loose with his torpedoes at whatever he finds on the surface. Finally, after all these weeks of crippling setbacks, he's ready to blow up some ships.

Lange peers through the periscope. *Let's see what we have here.* The periscope breaches the surface.

Oh, mein Gott!

"Destroyer!"

Lange swings the periscope around and scans the surface southwest.

Another destroyer! And another!

Ficken!

Three enemy destroyers! And what is that in the distance? An aircraft carrier! And flying above it—fighter jets!

Lange retracts the periscope.

Tauchen! he calls out. "Dive!"

The Americans are almost upon him.

TWENTY-THREE

"AWAY, ALL BOARDING PARTIES!"

Captain Gallery has had enough.

All day Saturday, June 3, 1944, his fleet of fighters and torpedo planes are doubling back on the same search area from the night before. And they are still coming up empty—a "complete blank." Nothing is out there. Right now he's positioned a hundred miles off Cape Blanco, a peninsula between Mauritania and Western Sahara on the African Coast.

June 4.

Sunrise.

The aircraft carrier is awakening. It's exceptionally warm, with a gentle breeze blowing across the flight deck. Commander Trosino reports to Gallery on the bridge with his fuel supply update. Trosino looks beaten.

"You better pray hard at Mass this morning, Cap'n. You used more oil than I figured on last night."

Gallery has to concede it's time to call it quits. He gave it his best shot, but he must abandon his quest to capture a U-boat on the high seas. He orders Task Group 22.3 to set course for Casablanca, fifteen hundred miles north. He regrets not having listened to Trosino earlier, because now he doesn't know if he'll have enough fuel to make it. Acute embarrassment and peril await him should he require a tow. It's a fiasco.

Reveille is sounded and the half-asleep crewmen roll out of their racks and stumble their way to chow. Nobody wants to miss Sunday chow, which is generally the best meal of the week.

At daybreak, general quarters is sounded.

"All hands—battle stations."

Big yawn. Everyone knows it's a routine drill.

At 7:00 a.m., the boatswain's mate blows his whistle to announce morning prayers. The chaplain, Father Weldon, recites Mass in the library. At 8:00 a.m., the crew "turns to"—meaning they get to work. At 9:00 a.m., it's the flight deck parade as a crewman nicknamed "Muscles" leads morning calisthenics.

"One-two-three-four. One-two-three-four."

High Mass follows at ten fifteen. All hands on the hangar deck stop what they are doing, doff their caps, and pay attention. Smoking is forbidden. All card games are put aside. Radios and phonographs are turned off as Father Weldon delivers a thirty-second devotion over the loudspeaker.

Gallery is in his cabin when his orderly hands him a mimeographed sheet. It's the *Guadalcanal*'s Plan of the Day, setting forth operational instructions for June 4, 1944, and distributed to every man on the *Guadalcanal* and the other five destroyer escorts making up Task Group 22.3. One item in particular catches Gallery's attention: "Final Crew for Captured U-boat," followed by a list of more than two dozen names.

Gallery exhales with regret. For the last three weeks, the battle plan has been refined and the eager-beaver volunteers making up the boarding parties on the *Pillsbury, Chatelain, Flaherty, Pope, Jenks,* and *Guadalcanal* have all been training for the big moment and the breakout of hostilities. Gallery had been set on selecting sailors who had experience on submarines. Unfortunately, of the three thousand men on the six ships in the task group under his command, only one sailor had served on a US submarine, and that was as a yeoman, or clerk. In other words, an expert in filing systems and paperwork, a pretty useless skill when the seizure of a U-boat requires brute force.

Gallery is wistful. In very short order, the Plan of the Day will be posted on all bulletin boards on the *Guadalcanal* as well as the five destroyer escorts. The hunt for a Nazi sub is being abandoned and everyone will know it. He wonders if he'll be the target of snarky commentary in the berthing compartments when his lads gossip about the day's events, as they do every night. *Yeah, another cockamamie idea from Cap'n Gallery. Told ya it was loony from the get-go.*

Gallery is back on the bridge in the skipper's chair, with a bird's-eye view of the flight deck. A perfect day with clear skies, gentle breezes, and "soft sea." Only two fighter planes are airborne. It's a token patrol. What's the point? Gallery figures that U-boat he's hell bent on seizing is out of range by now, heading in the opposite direction. Gallery is still fuming over how that damn submarine got away.

It's time for Sunday lunch. On the menu today—chicken. The sailors jostle for position on the chow line because the choicest chicken parts go first.

Suddenly, general quarters is sounded.

"Man your battle stations!"

Jeez, another drill, they gripe. *On a Sunday?*

Nope, this is for real.

A submarine has been sighted.

■ ■ ■

When Ray Watts was in fourth grade in the tiny Texas panhandle town of Phillips, population under four thousand, he was given a test assessing how well he could hear. Turns out, he had the ears of a bat, scoring the highest grade in the school. Years later, in 1943, he was drafted and somehow the Navy knew all about that long-forgotten test—probably from the local draft board. Watts was sent to boot camp in San Diego and assigned to sonar school, learning all about anti-submarine sonar technology and equipment.

Now Ray Watts is on board the USS *Chatelain* as a soundman, sitting in front of a stack of screens in the sonar hut and doing what the Navy has trained him to do, which is to listen. His backup is standing behind him. He's a sailor named Chapman, who was drafted in 1943 and sent to Fleet Sonar School in Key West after the Navy discovered he was a musician and had a good ear.

Watts and Chapman make a solid team. They switch off every thirty minutes to stay fresh. Any longer and you run the risk of becoming tone deaf or bored by the monotony.

Watts is skilled at distinguishing pings. He can tell the difference between a school of fish ("mushy echo") and a submarine. Whales are a little more complicated. Their pings sound like a submarine, but not as sharp. Of course, mistakes happen. You can be teased unmercifully if you keep hitting the panic button. You never want to trigger general quarters because of a whale.

It is 11:10 a.m.

Watts hears a ping. And it's really sharp. So sharp it startles him. For a split second, he wonders if one of the other destroyers has crossed the bow of the *Chatelain* and that's her ping he's picking up. Watts leaps from his chair and looks out. The sonar hut is positioned right next to the bridge. Watts has the best seat on the ship. All he sees is beautiful blue ocean.

The officer of the deck is right there. "What's the matter, Watts?"

"I'm just seeing if we have a sub," he says. There's something out there. Maybe two thousand yards away. Maybe three thousand.

The ping is getting sharper.

"Contact 080," he calls out. Meaning eight hundred yards.

That's all the duty officer needs to hear. He informs the skipper, Commander Dudley Knox. General quarters is sounded.

Knox is a pipe-smoking, thirty-five-year-old commander who comes from a distinguished military family. One grandfather attained the rank of rear admiral and the other grandfather was an Army colonel. An aunt is the widow of General Douglas MacArthur's brother, Arthur MacArthur III. Knox was working as a reporter for the *Washington Herald* and living in a small townhouse in Georgetown with his wife, Lalla, and their two young daughters when he joined the Navy in 1939 following President Roosevelt's declaration of a national emergency. He committed to signing up for the duration of the crisis, plus six months. Knox was first assigned to naval intelligence as a lieutenant junior grade before his promotion to executive officer of *Chatelain*, then full commander. His family nickname is "Bangkok."

Captain Gallery must be informed.

Gallery is codename Bluejay. Knox is codename Frenchy.

On the *Guadalcanal*, the squawk box blares awake.

"Frenchy to Bluejay—I have a possible sound contact."

Possible sound contact? Well, nothing to get worked up about yet. Could be a whale. Or a porpoise. Sometimes a layer of cold water or other natural phenomenon emits a false sonar reading. Still, you never know. It can be exhausting and frustrating, but every sound contact must be treated like the real deal.

"Left full rudder," Gallery barks. "Engines ahead full speed." Then he grabs the Talk Between Ships microphone. "Bluejay to Dagwood—take two DEs and assist Frenchy. I'll maneuver to keep clear."

Dagwood is codename for Frederick S. Hall, commanding officer of the USS *Jenks*.

The task group immediately sets about executing Gallery's orders.

Following the Navy's rules of engagement, the *Guadalcanal* veers off, steering clear from the action. An aircraft carrier must be protected at all costs. As Gallery likes to put it, an aircraft carrier during an engagement with the enemy is like "an old lady in the middle of a bar room brawl—she has no business being there." Right now it's his job to get the *Guadalcanal* the hell out of there and offer his five destroyer escorts some elbow room to do what they're built to do.

As he swings west, Gallery sends a Wildcat fighter pilot—Lieutenant Wolffe "Bob" Roberts and his wingman, Ensign John "Jack" Cadle, flying an Avenger torpedo plane—into the air.

"Use no big stuff if the sub surfaces," he tells the pilots. "Chase the crew overboard with .50-caliber fire." It's a reminder meant for everyone: If at all possible, capture, do not destroy, that U-boat.

The *Chatelain* is still trying to confirm her submarine radar sounding. *Chatelain* is so directly on top of the mysterious object beneath her that the soundman still can't figure out what the pinging means. Just then, a solid reading from Watts. It's a submarine. Definitely a submarine.

It's happening. Dan Gallery's wild mission is actually underway.

Dudley Knox gets on the TBS box.

"Contact evaluated as sub—starting attack."

Knox wheels *Chatelain* around full rudder and maneuvers into attack position. In this manner, the ship makes a complete circle, pinging the sub with radar the entire way to double-check the sub's position. When *Chatelain* straightens out, she lets loose with a salvo of twenty hedgehog mortar shells. The hedgehogs are flung forward by a launcher, arching into the air, then splashing into the ocean. Unlike depth charges, which detonate by preset barometric fuses, a hedgehog requires direct contact with the submarine hull to explode. Each hedgehog weighs sixty-five pounds. You know it's a direct hit when you see an underwater explosion.

They all miss.

Pillsbury and *Jenks* are circling cautiously. Their sonars are pinging, just like *Chatelain*'s. This could be a German trap, luring the Allied ships into a circle.

Now the two fighter planes Gallery sent aloft from *Guadalcanal* see something.

"Sighted sub!" Lieutenant Wolffe Roberts radios that he can see the blurred shape of the enemy vessel submerged just below the surface. No way it's a whale, at that size. Both planes open fire.

"I put a shot right where he is," Roberts exults from his Wildcat. "I'll put down another burst. Destroyers—head for the spot where we are shooting!" the pilot screams.

A burst of .50-caliber bullets peppers the water.

"I could see him at twenty-five-hundred feet! Just put down another burst. At first, I could see him clearly. He is fading now. He is going deeper."

The commander of the *Pillsbury* wants to make sure he heard right. "Did you actually sight sub and fire?"

"Affirmative," Roberts responds. "I'll circle spot. Haven't seen him since last burst. Let's get the bastard. I wish I had ten thousand rounds."

Chatelain lets loose with a spread of twelve depth charges, set to trigger shallow.

Moments later, the depth charges all go off in a circular carpet of explosions. Gallery watches the ocean boiling with geysers of great white plumes.

All at once, the eruptions subside. The ocean is tranquil.

Eleven nail-biting minutes have passed since *Chatelain* reported the first sonar contact, and suddenly here come the words they've all been waiting to hear.

"You've struck oil, Frenchy. Sub is surfacing."

It's Ensign Jack Cadle, flying his Wildcat, informing *Chatelain* Commander Dudley Knox—codename Frenchy—that he's hit a gusher.

U-505 rises from the depths like a great gray whale. Ripples of white water pour off her sides. She is about a hundred yards from *Chatelain*.

But something strange is happening—the sub is running in a tight starboard ring. Her rudder is either jammed or she's positioning herself to fire a spread of torpedoes. There is no way of knowing.

In the *Chatelain*'s sonar hut, thirty-one-year-old radarman Joseph Villanella from Newark, New Jersey, is sweating bullets. He's pretty sure that the submarine is about to launch a torpedo with his name on it. His ship is broadside of the sub—a sitting duck. Villanella braces himself, wondering why *Chatelain*'s captain doesn't turn to make the destroyer a smaller target. The sub is pointing right at them.

"Oh, man," he says to no one in particular. Right then he makes a vow to himself: If he gets out of this alive, "I'm gonna go to church every Sunday."

On the *Guadalcanal* a torpedo pilot named Ritzdorf is also watching the action, and he can't quite believe it either. A Nazi sub? And why is it circling like crazy?

Ritzdorf races down to the ready room. There are eight pilots all laid back and relaxing, doing not much at all, lulled by those long days of inactivity. Ritzdorf is so worked up he's having trouble articulating what's taking place topside, while they're down here twiddling their thumbs.

His words come out in a burst: "Hey, you guys, there's a sub out there."

Ritzdorf has a reputation as a practical jokester. Because he is of German ancestry, he once thought it would be a goof to grow a mustache like Adolf Hitler. It got a lot of laughs, until he entered cadet training school in Seattle, where he was strongly advised that, with a name like Ritzdorf, he'd better shave that upper lip.

The pilots scoff at the jokester's words. "Go soak your head, Ritzdorf."

"Go back to bed," suggests another pilot who's dead-set against getting off that couch.

"No kidding!" Ritzdorf insists. "There's a sub out there!"

At last, they all decide to check it out for themselves.

Holy moly. He isn't kidding. There really is a U-boat, and she's spinning around like a gyroscope.

The news spreads from stem to stern. One pilot from Texas is so jazzed with adrenaline he starts pulling off his shoes and socks, ready to jump over the side and swim to the submarine—to do what, nobody knows—before the other pilots knock some sense into him.

For the next two minutes, the three American destroyers closest to U-505—Chatelain, Jenks, and Pillsbury—let fly with every small-arms weaponry they possess. U-505 is getting hammered with dozens of slugs fired from the arsenal of .50-, .40-, and .20-caliber weapons. The Wildcat and Avenger hovering above add to the cascade of lead blasting the vessel. Not to sink her, per Gallery's instructions. Just to hound the Germans into surrendering.

Peering through his binoculars, Gallery can make out the tiny figures of Germans popping out of the conning tower and hurling themselves over the side of the U-boat. They are leaping for their lives amid the relentless hail of bullets ricocheting off the submarine and peppering the ocean surrounding them.

Wait! What's that?

It's the wake of a torpedo, but it's not coming from the U-boat. It's been fired by the Chatelain. What the hell is Dudley Knox up to?

Turns out that Knox is determined to take out that sub. From what Knox can see, U-505 is swinging around to position her torpedo tubes to bear. Jammed rudder? Maybe. But who the hell can say for sure? As far as Knox is concerned, that boat is a dangerous wounded animal. He orders the *Chatelain* torpedo room to open fire.

There goes the torpedo. Gallery can see it. *Damn!* More and more of the German sailors are jumping overboard. The plan is working, damn it, but if that torpedo clobbers the boat, all his aspirations over the last three years to capture an enemy sub will go up in smoke.

But wait! Now he sees the *Chatelain*'s torpedo going wide by seventy-five yards and whizzing past *U-505*. Most likely Gallery never thought he'd bless the day one of his torpedoes misses its quarry. He is boiling mad when he gets on the TBS to remind everyone what this operation is about.

"I would like to capture that bastard, if possible," he growls.

The message is meant for all his commanders, but especially Dudley Knox.

11:26 a.m.

Sixteen minutes since the first ping reported by Ray Watts, the *Chatelain* soundman.

"Cease firing!"

The order is coming from Commander Frederick Hall on the *Jenks*. "Men are abandoning ship. There are lots of men in water." They're raising their hands, while clinging to their life rafts. The Germans are surrendering.

"All the men are off topside?" Gallery asks.

"I believe so. Not a damn one left."

"Do you think we can capture this guy?" Gallery asks.

It's game on.

What comes next is an ancient naval battle cry not heard on an American battleship since the year 1815 when the sloop-of-war USS *Peacock* opened fire and boarded the British brigantine HMS *Nautilus*.

"Away, all boarding parties!"

On the bridge of the USS *Pillsbury*, Boatswain of the Watch Wayne Pickels has watched the battle play out as the U-boat got pounded with small-caliber fire. He saw the German sailors leaping overboard and the absolutely bizarre spectacle of the stricken U-boat spinning around in a tight circle. He's rehearsed this drill for the last three weeks and knows what's coming next. Over the loudspeaker, he hears the voice of *Pillsbury*'s skipper, Lieutenant Commander George W. Casselman:

"Away, all boarding parties!"

Pickels relinquishes his boatswain's post and runs to the whaleboat where he is to link up with the other nine members of the boarding party.

Way down below deck, in the engine room, Zeke Lukosius hears the command and hurries topside.

Signalman Second Class Gordon Hohne also hustles over to the whaleboat with a flare gun tucked inside his belt and two sidearm .45-caliber pistols flapping against his hips. He's also carrying a portable signal light to communicate, if need be, by Morse code.

Lieutenant Albert David hears the cry and rushes from his post in the aft engineer room where he is assistant engineer and electrical officer. In no time flat, he's at the whaleboat.

One by one the other volunteers also report for duty: Arthur Knispel. Stanley Wdowiak. Chester Mocarski. George Jacobson. William Riendeau.

They're all a bit bewildered. A truly monumental task lies head, and they can't help but wonder:

Is this really happening?

THE NINE:
The boarding party that seized *U-505* on June 4, 1944, and made naval history. The original photo taken by the Navy mistakenly failed to include Lt. David. He was inserted only years later. Left to right: Chester Mocarski; William Riendeau; George Jacobson; Zenon "Zeke" Lukosius; Gordon "Flags" Hohne; Wayne "Mack" Pickels; Lt. Albert David; Stanley "Stash" Wdowiak; Arthur Knispel. *Courtesy of the Museum of Science and Industry, Chicago.*

Lt. Commander George Cassleman, perched on the bridge of the USS *Pillsbury*, his hands clasped over his head—a traditional Navy gesture of farewell to the boarding party on their way to seize *U-505*. Apparently Cassleman thought there was a decent chance he'd never see his men again. *Courtesy of the US Navy.*

The whaleboat—below right in photo—with Lt. David's boarding party from the USS *Pillsbury*, approaching *U-505* for the first time. *Courtesy of the National Archives.*

Sailors on the US Aircraft Carrier *Guadalcanal* watch as *U-505* comes under tow. Note presence of the boarding party, gathered on the submarine's bow. *Courtesy of the National Archives.*

U-505 seized, but barely afloat, bow jutting out while the rest of the boat is submerged, with just the conning tower and the American flag above the surface. *Courtesy of the Museum of Science and Industry, Chicago.*

American sailors on the conning tower of *U-505*, barely above the surface while the rest of the submarine is submerged. *Courtesy of the Museum of Science and Industry, Chicago.*

ABOVE:

Captain Gallery, on the conning tower of *U-505*. Note position of US flag over smaller Nazi swastika, a symbol of victor over vanquished. Note: damage to the conning tower from American firepower. *Courtesy of the National Archives.*

RIGHT:

Cheeky Captain Gallery—always looking for a good laugh—on the flight deck of the *Guadalcanal* wearing the confiscated cap of German *U-505* skipper Harald Lange. Photo was taken June 10, 1944—six days after the capture—during the tow to Bermuda. *Courtesy of the National Archives.*

Captain Dan Gallery and Lt. Albert David on board the USS *Guadalcanal*, following the capture of *U-505*. *Courtesy of the Museum of Science and Industry, Chicago.*

ABOVE:
German POWs. Photo taken June 16, 1944, on the USS *Guadalcanal*. Hans Goebeler is second from the right. *Courtesy of the Museum of Science and Industry, Chicago.*

RIGHT:
Just some of the many packs of cigarettes found on *U-505*. *Courtesy of the Museum of Science and Industry, Chicago.*

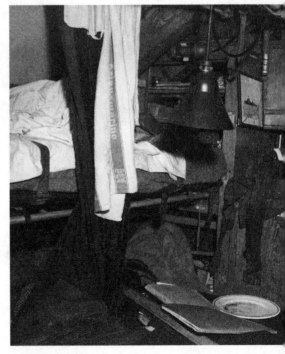

ABOVE:
Wounded German sailor after the battle on June 4, 1944, being taken into American custody. *Courtesy of the Museum of Science and Industry, Chicago.*

RIGHT:
The skipper's cabin on *U-505*, taken by a Navy photographer the day of *U-505*'s capture. Note the empty plate. Kapitanleutnant Harald Lange was finishing his lunch when the submarine came under attack. *Courtesy of the National Archives.*

US Navy Commander Earl Trosino on board *U-505*. His engineering know-how proved vital in saving the boat. In this blurry photo shot aboard the rolling and pitching vessel, he has taken on the miserable task of emptying the boat's "shit bucket"—because no one else would volunteer. Note Trosino's Panama hat. He found it on board the submarine after his US Navy hat washed overboard when he "crash-landed" on the deck of *U-505*. *Courtesy of the Museum of Science and Industry, Chicago.*

ABOVE:
The Enigma machine found on the submarine. *Courtesy of the Museum of Science and Industry, Chicago.*

LEFT:
The *U-505* galley. Large enough for just one cook. *Courtesy of the Museum of Science and Industry, Chicago.*

HEADQUARTERS OF THE COMMANDER IN CHIEF

UNITED STATES FLEET

TOP-SECRET   MEMORANDUM

Date: **7 June 1944**

From: Assistant Chief of Staff (Anti-Submarine)

To: Commander, TENTH Fleet

1. Under the assumption that U-505 is successfully towed to Bermuda, it is suggested that she be given an appropriate cover from which she should also carry, if commissioned.

2. Two suggestions for your consideration are:

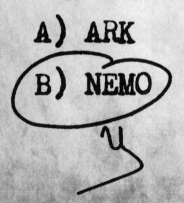

A) ARK

B) NEMO

Very resp'y,

F.S.

TOP SECRET

I fully realize that should Germany learn from any source of the capture of the U-505 that the tremendous advantage which we have gained by her capture would be immediately nullified, resulting in a great loss to the United States.

I am also aware that should I break this oath, I will be committing a grave military offense, thereby subjecting myself to a general court martial, which court has the authority in time of war to award as punishment the death sentence.

LEFT:

Memo proposing two codenames for *U-505*: "Ark" and "Nemo." The ultimate selection was made by Fleet Admiral Ernest King. *Courtesy of the Museum of Science and Industry, Chicago.*

ABOVE:

Oath every sailor on Task Group 22.3 was ordered to sign, swearing them to secrecy regarding the capture of *U-505*—under penalty of death. *Courtesy of the Museum of Science and Industry, Chicago.*

Earl Trosino where he felt most at home—in the machine shop working on engines. *Courtesy of the Trosino family.*

Wayne "Mack" Pickels Jr., 1942, shortly after he surprised his family by enlisting in the Navy instead of the Army. *Courtesy of the Pickels family.*

Albert Leroy David and his wife, Lynda, a nurse. They met at the University of Wisconsin, where the Navy had sent him to study diesel engines. *Courtesy of the Museum of Science and Industry, Chicago.*

The sailor in white standing precariously on the conning tower is Master-at-Arms Leon "Ski" Bednarczyk. His medal citation reads: "Through his knowledge of foreign languages and through his brilliant intelligence work among the captured Nazis, he obtained technical information without which the submarine probably could not have been saved." *Courtesy of the US Navy.*

TWENTY-FOUR

ABANDON SHIP

U-505 is in real trouble—the worse this submarine has ever known. Harald Lange orders a single torpedo fired in the general direction of the enemy ships but there's no time to calculate for distance, firing angle, or velocity. There's very little chance of hitting a target except by fluke, even with the acoustic torpedo. It's a blind shot, a Hail Mary—a diversionary tactic to distract the Americans.

Inside *U-505*, there is chaos. Lange orders a crash dive to maximum depth.

A strange metallic banging or "clinking" is coming from above. Hans Goebeler stares at his fellow sailors, but they all look as perplexed as he is. In their combined years of service in the Kriegsmarine, they have experienced every sound there is to hear in the peculiar undersea world of submarines, but never anything like this. What the hell is that? Could it be a heavy chain dragging across the hull of the boat? Have they snagged a sea mine that will explode at any second? Then they grasp that it's the plinking of .50-caliber bullets striking the hull, fired from the air by American planes. The ocean water is so crystal clear the pilots can make out the outline of *U-505* even at this depth.

Lange orders a deeper dive, and this is when another surprise wallops them. It's a full salvo of hedgehogs. They await destruction but by sheer serendipity not one mortar from the cluster of rockets hits the U-boat.

Lange orders a sequence of gyrating seesaw maneuvers.

Fifty meters down, depth charges explode around them. *U-505* shudders violently and all the lights go out. Machinist Hans Decker hears something that every submariner dreads—the sound of water flooding in.

"Rudder taken out—water breach!" somebody screams.

Major flooding is declared in the aft torpedo room, and Lange orders it evacuated. The last man out, Wolfgang Schiller, slams the watertight hatch behind him.

Schiller is wearing Navy shorts, a civilian shirt, canvas shoes, and no socks. As he makes a dash for the control room, he comes across a lamb's wool sweater just lying there. It must belong to one of the officers. Well, finders keepers. Schiller puts it on. You never know how long you'll be in the water. It could turn cold out there.

Flashlights come out. In the pitch-black engine room, Hans Decker shines his torch on the tangled mess of broken dials and pipes.

Lange absorbs all the damage reports coming his way. The main rudder is jammed, and the boat is out of control, turning in a clenched circle. It's a runaway dive, down to 230 meters, and it's only minutes before the pressure hull reaches the depth at which she will be crushed like a walnut. Without the ability to steer, there is no hope. The boat is lost.

"Take us up," Lange cries out. "Take us up before it's too late!"

The ballast tanks don't respond and the diving planes, which jut out of the U-boat like wings, are jammed downward. *U-505* is sinking to the bottom of the ocean.

Then, somehow, the compressed air being pumped into the ballast tanks start blowing out the water, and the next thing the crew knows, they're on a vertical climb upward.

The sailors stare at one another, their faces drenched with sweat, eyes wide with terror. What now? As they ascend to the surface, they all grasp the fact that Lange has no option other than to issue the orders to abandon ship.

Thirty seconds pass.

Sixty seconds.

Still, Lange is silent. They are seconds away from breaking surface.

Of course, there is another option. A fanatical Nazi skipper might attempt to take down one of those American destroyers, or even the aircraft carrier. What a glorious way to die for the Fatherland! But Lange assesses the situation nautically, not zealously. His boat is spinning uncontrollably. He is without a rudder, with no access to the torpedoes. There is nothing to do but surrender.

U-505 breaches the surfaces and is met by a barrage of bullets blasting into the conning tower. That does it.

Raus schnell!—"Abandon ship!"

The men assemble on the ladder. Most of them are wearing shorts. They each carry a life preserver that fits over their heads and under the crotch with an adjustable strap and is inflated by a carbon dioxide cartridge.

Lange pushes them aside. As skipper, it is his duty to peer out the hatch before anyone else and evaluate the tight spot they face with his own eyes, come what may. Hearing the sharp pangs of the American bullets raining down on *U-505*, Lange must be thinking he'll be cut down for sure in a hail of gunfire. He is between the devil and the deep blue sea.

Lange pops open the hatch and steps onto the bridge, knowing he could be drawing his final breath.

Right behind him is his executive officer, Paul Meyer.

Behind Meyer are the watch officers.

Behind the watch officers—everyone else.

Still wearing his black silk pajamas, Lange squints against the bright sunlight as a plane swoops down and strafes the deck of *U-505* with machine-gun fire. Lange takes bullets in his legs and knees. It feels as if he's been hit by lightning. A wood splinter impales his right eyelid—he's blinded in one eye.

Meyer is also hit, sustaining a serious scalp wound. He's sprawled across the starboard side, blood streaming down his face.

Half-blind and delirious with pain, Lange shouts down at his crew.

"Every man—off the boat! Sink the boat!"

With those last orders, the Old Man loses consciousness.

Scuttling the submarine is the standard order of business for any well-disciplined U-boat crew. Like the US Navy's sacred battle cry "Don't give up the ship," the Germans, above all, must keep their vessel from falling into enemy hands.

Only three officers have the security clearance and technical know-how to set the fourteen demolition charges scattered throughout *U-505*: Lange, First Officer Meyer, and Chief Engineer Josef Hauser. The demolition charges are small, difficult to locate, and dangerous to set if you don't know what you're doing. They are tiny bombs and notoriously unreliable due to seawater corrosion. They are like forgotten fire extinguishers in a tenement building, rarely serviced or checked for dependability if anyone can even remember where they are concealed.

With Lange and Meyer wounded in action and out of commission, the scuttling duty is left to the officer most detested by the crew—Hauser, the "Raccoon."

Wolfgang Schiller, still clad in that purloined lambswool sweater, stampedes up the ladder. He takes a moment for his eyesight to adjust to the bright sun. Then, with a big splash, he makes a headlong dive into the water. When he surfaces, he swims as fast and as far from the boat as his strokes can take him.

Bullets whiz by as Petty Officer Karl Springer pokes his head out of the conning tower hatch. Gottfried Fischer, the radio operator, is right there, slumped across the top deck. Fischer is affectionately known as Gogi. He is one of the few crew members who is married.

"Gogi, come on!"

Then Springer realizes Fischer has been shot to pieces.

Hans Goebeler awaits his turn up the ladder when some inner voice tells him to step aside and stay with the boat. *Where is Hauser?* He was just here. *What happened to the chief engineer?* Goebeler thinks, *The gutless bastard.* Hauser must have evacuated *U-505* to save his own neck.

In Hauser's absence, Petty Officer Alfred Holdenried tries to organize a scuttling plan of action. He orders the men who are still with him to open the forward main diving tanks, #6 and #7, but the valves won't budge. The operating shaft must have been twisted by the depth charge explosions. That's the end of that. Even Goebeler realizes the time has come to get off the boat.

Wait one second. An idea hits Goebeler as he starts climbing the ladder. The sea strainer! Goebeler leaps off the ladder and runs back to his duty station near the control room. There it is. The sea strainer. It's a ten-inch-wide filtration system, the size of a bucket. All Goebeler has to do is unlock the four clamps that hold the strainer in place. He does that, and suddenly a gush of water gurgles up. Goebeler has effectively punctured a hole in *U-505.* He watches as the ocean water comes in as a flood, spurting like a geyser. That should do it. The strainer is situated in the lowest end of the boat and *U-505* will surely sink to the bottom now. How bittersweet for the sailor who cherished *U-505* perhaps more than any other crewman to be sealing her fate as her executioner. Goebeler tosses the sea strainer cover into a corner of the control room and runs back to the ladder and evacuates. So far as he can tell, he is the last man on board, the last to say goodbye to *U-505.*

When he hoists himself out of the conning tower hatch, Goebeler realizes what *U-505* has been up against. He's looking at the full power of the United States Navy—five destroyers and in the distance an aircraft carrier, dazzling in the morning sun.

Goebeler catches sight of a pack of his fellow Germans, bobbing around in the water. He crawls along the top deck, now slimy with brine and sticky blood, and slides off the submarine, splashing into the ocean. While the temperature is agreeable, the water is choppy. Goebeler swims toward a large rubber life raft filled with his comrades. Then he climbs on board, and they start paddling away.

Goebeler looks back for a final glimpse at *U-505* before the boat goes under. *Hmm.* She's still afloat, circling wildly in a counterclockwise rotation. Not to worry. She'll go down soon. He made sure of that when he unplugged the sea strainer, which is as effective as any demolition charge, only a little more sluggish. It's just a matter of time.

More Germans swim to the raft and climb aboard, and soon it's filled to capacity. Petty Officer Springer is worried that the raft is overcrowded. It's sitting too low in the water, and each ocean wave brings another swell of water. They need something to bail with. *Does anyone know what happened to the skipper?* Word is he's alive on another raft but critically injured. They've all seen Gottfried Fischer's body. Poor Gogi.

Wolfgang Schiller is adrift in the ocean. One American destroyer swings by and he's tossed a rope, which Schiller can't catch. He's nearly chopped to pieces when the propeller comes within spitting distance. Then he makes out Goebeler's life raft and swims for it.

What is that?

Something just brushed Schiller's leg.

"Is there a shark?"

"*Nein, nein,*" Goebeler calls out.

Schiller isn't so sure. The ocean is stained with the blood of the wounded and that is certain to attract man-eaters in these shark-infested waters.

There it is again. "Is there a shark behind me?" Schiller is certain there's a gray shadow nipping at his heels.

"You're seeing things," Goebeler calls out.

Then Schiller realizes what it is. In his haste to escape from *U-505*, he neglected to tighten the straps of his life preserver. The loose strap is flapping against his bare calves.

And now they all await the arrival of their conquerors. Once again, Hans Goebeler looks over his shoulder to check on the condition of *U-505*. The submarine remains on the surface, circling widely, stubbornly refusing to sink, but definitely sitting lower in the water. What a tenacious boat.

Then Goebeler is stunned to see a compact American motorboat bouncing across the water, making its way in the direction of *U-505*. What are their intentions, he wonders. And who would be foolhardy enough to board a sinking submarine?

TWENTY-FIVE

HI-YO, SILVER!

At 11:38 a.m., the whaleboat is launched over the side of the USS *Pillsbury* and lowered into the Atlantic. On board are the nine sailors from the boarding party, ready to take a ride into the history books.

Lieutenant Albert David looks back at the *Pillsbury*. There's Lieutenant Commander George Casselman, perched on the bridge staring back at him—and Casselman's hands are clasped over his head. As a veteran familiar with Navy lore, David knows what the gesture symbolizes. It is the Navy's traditional silent farewell. Apparently Casselman isn't expecting them to return from the mission. David turns to his men.

"Well, fellows," he says, "we may not be back, but we'll give them a hell of a fight."

At the wheel of the twenty-five-horsepower whaleboat is Coxswain Philip Trusheim, from Huntington Beach, California. Trusheim enlisted in the Navy in 1941, just before Pearl Harbor. He was seventeen. He's got plenty of experience operating the whaleboat, saving the lives of downed pilots, going on routine mail runs, or ferrying passengers from shore to the *Pillsbury* and vice versa. He's highly skilled at piloting the nimble twenty-two-seat craft, but he's never done anything like this. Hell, *nobody* has. The U-boat is about two thousand yards up ahead. Crazy thing is, the submarine is circling counterclockwise at six, maybe seven knots. Sure is going to complicate things.

Trusheim is skimming along the surface when they encounter a cluster of German sailors in life jackets afloat in the water.

"Save me, comrades," one German beseeches. "Save me."

Trusheim ignores them. He doesn't want to see anyone drown, but he knows that other whaleboats are on their way and these Germans will be rescued soon enough. Trusheim has to focus on the mission, which is to deliver the nine men from the boarding party to that sub as quickly as possible. And the clock is ticking.

In the whaleboat, Zeke Lukosius feels pity for the Germans in the water. *There but for the grace of God go I,* he thinks. "Somebody's gonna pick you up," he assures them as he barrels past them. "We're not going to abandon your asses." He also can't help sticking it to the Germans: "We're going after the submarine."

Everyone else is tight-lipped as they cross the water. No conversation at all, each man lost in his own thoughts.

It takes Trusheim nine minutes to reach *U-505.* Now he's got to figure out how to get the boys on board that swiveling submarine. Trusheim has never seen anything as screwy as this. He figures the absolute dumbest thing he can do is chase after the sub and try to catch up as she's swinging about in a circle. Instead, he cuts across at a collision course to the axis of the swinging. It's a gutsy maneuver and it works. Trusheim pulls alongside the sub's starboard side. He's right by the conning tower, sticking close like a pilot fish to a shark.

Now it's all up to Seaman First Class Jim Beaver Jr., who is brandishing the whaleboat's bow hook.

Beaver grew up in the pipsqueak village of Atco, Georgia. Maybe two hundred folks live there. And here he is, about to make a leap that, he has been informed, could contribute to bringing about an end to World War II. Beaver leans forward and snags the bow hook to the submarine. Then he bounds out of the whaleboat and lands on the top deck of *U-505,* riding it like a bucking bronco. The kid from Atco is officially the first American to board an enemy sub on the high seas in time of war.

From the bridge of the USS *Guadalcanal,* Captain Dan Gallery observes the gallant lad leaping on to the slippery and bloody deck of the U-boat, and he is in awe. Beaver reminds him of that enduring American legend, the Lone Ranger.

Gallery is so worked up, he flicks on the TBS.

"Hi-yo, Silver—ride 'em, cowboy!"

With Beaver's bow hook serving as an anchor, Trusheim swings the whaleboat over and snags the stern with another hook. The whaleboat is still bouncing around like crazy against the side of *U-505* but at least Trusheim and Beaver have successfully moored their rinky-dink vessel to the submarine.

On the *Guadalcanal,* Captain Gallery receives this update: "Our boat is alongside of submarine now. Please don't let planes fire anymore."

Gallery acknowledges the report. "Roger. Nice going."

Now it's time for the boarding party to show what they can do. Gunner's Mate First Class Chester Mocarski is the first to proceed. He's the brawny guy from Brooklyn who has been tasked with conveying the toolbox and a thirty-foot-long steel chain over to the submarine.

Lugging all that gear is challenging, even for a muscleman like Mocarski. As he makes the jump, bad luck strikes as the heaving whaleboat and *U-505* suddenly pull apart in the choppy sea. Then the two vessels slam into each other—with Mocarski wedged between. His legs are crushed, and several ribs are broken. Mocarski screams and plunges into the ocean, losing his grip on the toolbox and with it the steel chain. Pickels and Lukosius manage to haul Mocarski out of the water. The boarding party already has its first casualty. Not a great start.

Pickels watches in despair as the toolbox and chain go deep six into the ocean. It's a major hiccup. His heart is sinking, along with the equipment.

Nothing they can do now but carry on with the mission as best they can, minus a key player and vital equipment.

One by one they make the jump on to the pitching submarine. Arthur Knispel. Stanley Wdowiak. Wayne Pickels. Zeke Lukosius. George Jacobson. William Riendeau. Gordon Hohne. And of course their leader, Albert David.

They're all on board *U-505* now.

Hohne, a.k.a. "Flags," positions himself on the conning tower with his handheld Navy semaphore flags. He's dressed in whites to make himself stand out. He'll be the eyes and ears of the boarding party, communicating everything that's happening to Gallery and the commanders on the destroyer escorts.

Lieutenant David looks around. The body of the U-boat sailor Gottfried Fischer is sprawled across the deck, face down.

The hatch is open and David peers inside. It's the size of a sewer cover and so dark down there he can't make out a thing. Torpedoman Knispel, armed with a Tommy gun, is designated to be the first sailor to go down the hatch. He's prepared to shoot it out with any German he finds inside the sub. But David announces a switch. He is going in before anyone else. First to enter, first to face enemy gunfire.

He takes a deep gulp of air. "Well," he says, "let's get her over with."

This is it. He's been told the Germans have more than likely set demolition charges that will detonate in a matter of seconds. He also knows that simply by climbing down the ladder he could trigger a booby trap planted by the Nazis. Then there's the distinct possibility there are Germans still inside, ready to cut him to pieces the moment he enters the sub. He readies himself for hand-to-hand combat.

Here goes nothing. Whatever happens, it'll be over quickly.

Lieutenant Albert Leroy David proceeds down the hatch.

Right behind him are Knispel, the nineteen-year-old baker's son from Newark, and Radioman Third Class Stanley "Stash" Wdowiak, the kid from Brooklyn.

It's pitch-black inside. David and Knispel turn on their flashlights and aim the rays of light down the body of *U-505*, bracing to face machine-gun-toting Nazis who've stayed behind. Instead, they behold a dizzying array of valves and pipes and instruments. All they hear is the murmur of the electric motors churning, and the throbbing of the propellers. A steady hum, made haunting by the fact that there doesn't appear to be a soul on board except for them. It's as if they have stepped into the *Flying Dutchman*, the ghost ship of legend.

Suddenly, their nerves are jangled by a strange noise coming from the control room, a ticking sound. A time bomb? No, it turns out it's just the ticking of a gyroscope. Whew . . .

Then, another ominous sound: the gurgling of rushing water.

David yells out to let the rest of the boarding party know that the submarine is in their possession. Lukosius and Jacobson are next to climb down the ladder.

Locating the source of the rushing water is their first priority. *U-505* is tilting 10 degrees down by the stern. David wonders how much time they have left. They've got to plug up the leak or the submarine will sink for sure with all of them inside. That would be a first—American sailors, perishing aboard a sunken German U-boat.

Lukosius and Jacobson split up. Jacobson is the chief motor machinist's mate, and at age forty-four, the old salt of the boarding party. He heads for the engine room to shut off the electric motors. Lukosius proceeds in the direction of the control room.

It doesn't take Lukosius long. Just off the control room he ducks behind the main periscope and encounters the source of the leak—a spout of water shooting up five feet. It's the sea strainer that Hans Goebeler had unlatched to scuttle the submarine.

Lukosius looks around. The cover must be here somewhere. He explores some more. Wait. Over there. Is that it? What a lucky break! It's the right shape and size. Lukosius is familiar with sea strainers, or "sea cocks" as they are also called, because it's pretty much the same design on American ships. Submarines and surface ships all require a steady source of seawater to flush the toilets, cool the engines, and feed the freshwater distiller.

Something is rattling Lukosius. Maybe this *isn't* a lucky break. All the warnings about booby traps have put him on edge and he's beginning to wonder if finding the sea strainer cover is too good to be true. There could be a bomb hidden underneath. He inspects it for any signs of tampering or wiring. Will he trigger an explosion the moment he lifts the damn thing? Why would the Germans have just left it there?

Wayne Pickels now joins him. They stare at each other. If it's a booby trap, they're both goners, heading straight for Davy Jones's locker.

"Wayne," Lukosius says through tight teeth, "here goes nothin'."

Lukosius takes a deep breath and lifts the sea strain cover. *That's a relief.* No booby trap. No explosion. Lukosius works feverishly to clamp the cover in place against the rushing water and tighten all the butterfly nuts. Job done! No more water spout shooting five feet into the air. And in the nick of time. Another few minutes and the sub could have gone down, taking the entire boarding party with her. Lukosius shakes his head in disbelief. If that sea stainer cover had been kicked down the submarine just a little farther, or tossed into the bilge, he never would have found it. Somebody up there must like him.

All the men from the boarding party are now inside *U-505*, except for Flags Hohne and the stricken Chester Mocarski, who is left sprawled topside, out of action, and in agonizing pain. Hohne is translating communiqués from Captain Gallery. Mainly they're on the order of "What is going on now?" Gallery demands to know every detail.

Stash Wdowiak proceeds to the submarine's "radio shack." He steps inside the tiny cubbyhole filled with an array of radio equipment, and there he finds an intriguing wood box measuring 13.5 by 11.6 inches. Inside the box is a machine that looks like a futuristic typewriter. It's an Enigma encryption machine that scrambles ordinary words into gibberish. Stash has no idea how Enigma works. Somebody else is going to have to figure that out. But he knows that in his hands he's holding one of the great treasures of the war.

A human chain is formed. Stash passes the Enigma machine to Mack Pickels, who passes it to Zeke Lukosius, who passes it up the ladder and through the hatch to Hohne on the conning tower, who tosses it over to Jim Beaver in the whaleboat. It's a bucket brigade for enemy secrets of incalculable value. More equipment and documents follow. The whaleboat is getting so crammed and Beaver so harried, he wonders if he should jettison unimportant items to make room. Like this thing that looks like a cruddy typewriter. Maybe he ought to pitch it overboard. Then he reconsiders and finds the space.

It's the Enigma machine.

Down below in the submarine, the boarding party continues digging for documents and any equipment that looks sophisticated and technologically advanced.

Bingo! The German code books. They are the mother lode. The code books are printed with pale, red-colored ink that dissolves the instant it comes in contact with water. A warning in German is printed on the cover: *Vorsicht! Wasserloslicher Druck!* "Careful! Water Soluble Print!" The slightest spray of water and the ink bleeds. All the *U-505* crew had to do was dump a bucket of water over them. To further foolproof the code books from falling into enemy hands, the covers are made of lead, so they'd sink. Hard to believe the Germans could have left them behind, yet here they are, up for grabs. Another sign of the terror that sent the Nazi crew running.

In no time the radio room is stripped bare. The Americans work at a speedy pace because they know the sub could sink at any moment.

Meanwhile, Lieutenant David hastens for the captain's cabin.

It has a small desk, sink, bunk with a thin mattress and a portrait of Adolf Hitler—compulsory on all U-boats. Judging from the unwashed porcelain plate that David finds on the table, this captain must have just eaten his lunch. The bed is unmade, and clothes are strewn about.

David finds what he's looking for—the skipper's safe. He attacks it with a crowbar and the safe pops open with a sudden crackle that takes him aback. It sounds like an explosion. Then he digs in and stuffs all the paperwork he finds into a seabag. One more look around. A framed family photo on the captain's desk. Must be the captain's wife back in Germany. David scoops it up because you never know if it might come in handy.

Wayne Pickels passes the mess hall. He sees half-eaten trays of food left on the long table.

Pickels joins David in the captain's cabin to help out. He busts open the locked drawers and finds a trove of maps and documents and the submarine's official logbook. So, that's the name of the boat: *U-505*.

A terrifying find—a demolition charge, right at the base of the periscope! It weighs about five pounds and is triggered by a timer, but thankfully it hasn't been set.

Lieutenant David is now in the forward torpedo room looking for more demolition charges. He finds a bunch of them tucked between the torpedo tubes. Once again, they have not been activated. Unbelievable.

David grasps the wires. He has no idea what will happen if he pulls them out, but does he have a choice? Arthur Knispel is right behind him, and he looks like he's ready to jump out of his skin.

"Well," David says, "here goes nothing." That expression sure is making the rounds. David yanks—hard.

The next thing, he's holding a fistful of wires. There's a sheepish grin on his face. Then he turns to face Knispel, but realizes he's been talking to himself. Knispel's not there. Knispel had slipped out of there, and David doesn't blame him one bit.

Now David heads for the engine room to check on how Jacobson's doing. The chief motor machinist's mate is still evaluating the electrical motors and figuring out how to shut down the system. Naturally, everything is in German and Jacobson's really struggling with interpreting the meaning of the legends on the dials and levers and scared to touch any of them for fear of blowing the boat out of the water. He had rehearsed this moment by studying captured German diagrams, but nothing could really prepare him for the real thing.

From his position on the top deck, Flags Hohne yells into the hatch that he's just received a communiqué from the USS *Pillsbury*. They want the engines switched off pronto. They want the submarine to stop spinning.

"What the hell do you think we're doing!" David yells back.

David needs to catch his breath. He finds a wooden box with a leather pad for a seat and parks himself on it as he ponders his next step. What he doesn't know is that there's a demolition charge hidden on the underside.

Finally, Jacobson thinks he's got it figured out.

He flicks a switch. All at once, the steady hum of the electric motors goes silent. Success!

But not so fast. The sub stops spinning, yes, but it starts to seriously dip. By shutting off the engine, *U-505* has lost forward momentum. The sub is sinking.

Topside, ocean swells are washing over the conning tower.

"Water is coming in the hatch," Hohne hollers. "I'm gonna shut her up."

Next thing they hear is the watertight hatch being slammed. Hohne has stopped the flooding—but he's also locked the boarding party from any means of escape in an emergency.

A fighter pilot observing everything is giving Gallery a play-by-play account of what just happened.

"Roger," Gallery responds.

David tells Jacobson to flick the switch back on, pronto. The next moment, the electrical hum kicks in again and the engine is restored to life, halting the sub's descent.

That was close. David figures the boarding party has officially run out of time.

"The sub is sinking," he snaps to his men. "Everybody get topside!"

Hohne lifts the hatch open, and David orders him to send a message to *Pillsbury* via semaphore: They need a tow, right away.

In spite of the lieutenant barking at him, Wayne Pickels isn't anxious to go just yet. *Risk my life seizing a Nazi sub, and return without any souvenirs? Forget that!* He goes on a scavenger hunt and throws everything he can lay his hands on into his seabag.

Back in the captain's quarters, Pickels finds a nifty fountain pen. It's a Montblanc. Very nice. Also, a Nazi submarine insignia and a medal.

Then he darts into the crew quarters and finds a lot of really cool stuff. He collects five pairs of high-end Zeiss binoculars—some say they are the best in the world. Plus a Belgian Mauser revolver and a leather officer's coat. There's more. Pickels pops open the individual lockers with a crowbar and finds two bottles of brandy. He also rifles through the cabinets, which are filled with tin cans of food. He discovers a supply of aftershave lotion—what the boys on the *Pillsbury* would call Foo Foo water. Looking at the display, Pickels concludes the German sailors must have liked to spritz themselves with heavy doses of cologne when they went ashore to meet the ladies. He'll later learn that the bottles of scented water have nothing to do with dating French ladies of the night. The Germans sprayed themselves to cover up body odor because they couldn't take many showers during war patrols.

Into the seabag it all goes—including the tin cans, brandy, and Foo Foo water. Thinking ahead, he stuffs the Montblanc fountain pen and a Kreigsmarine sailor's cap into his pocket, just to have *something* in case his seabag is confiscated. Not a bad haul at all.

Up above, Flags Hohne's arms are weary from nonstop signaling, and he figures he's got one shot at seeing the interior of the submarine just once. He's been stuck on the conning tower all this time, and there's no way he isn't going down for a quick look-see. Plus, he's got to find himself a souvenir, just like the other guys.

Hohne climbs down the hatch and looks around. He has a working knowledge of German because his parents spoke it at home, but everything on *U-505* is written in military jargon and abbreviations, and he can't figure out a word. He finds a pair of binoculars, a Luger, and a long-barreled Mauser pistol. Hohne ends up in the mess hall. There's a beer on the table and a liverwurst sandwich. He takes a big bite. Still fresh.

He ate the Nazis' lunch.

TWENTY-SIX

"WHO'S IN CHARGE?"

Hans Goebeler is on the life raft swaying in the ocean, grateful to be alive. He was alarmed at first by the sight of that American boarding party, but now he watches with a smug and self-satisfied expression as the American boarding party chases after *U-505*, which is circling counterclockwise at a high speed. The scene recalls something out of the Keystone Cops. What a crazy sight.

The stern is completely submerged. Only the bow and the top of the conning tower are poking above the water. Goebeler informs the other sailors on his life raft that he left a parting gift for the Americans—he sabotaged the submarine by detaching the sea strainer cover. It's only a matter of time before *U-505* goes to the bottom.

"Three cheers for our sinking boat!"

All the Germans respond, "Hip hip hooray!"

Then Goebeler sees the American boarding party leap aboard the submarine and climb down the hatch. Good! His first thought is that when the boat sinks she'll take all those Yankees with her. But he has to grudgingly admit that he has never witnessed anything as brave as these US sailors who are venturing into an enemy submarine that they know is about to go under. Unbelievable.

It should happen at any second.

And yet . . .

U-505 refuses to succumb.

A sick feeling is starting to overwhelm Goebeler. Could it be? Have the Americans plugged the leak?

Now the USS *Chatelain* is pulling alongside the German raft to come to their rescue. Goebeler and the others clamber up the netting and find themselves on the deck of the destroyer, prisoners of war of the United States. And they are grinning ear to ear. They are alive! For them, the war is over.

The wounded are taken to sick bay. This includes Kapitänleutnant Harald Lange, whose right leg has been shattered by gunfire.

With Tommy guns pointed at them, the Germans are herded together, stripped, and searched. They are obliged to provide only their name, rank, serial number, and date of birth. Time will tell how long that lasts.

Goebeler is taken aback when an American snatches the St. Christopher medal off his neck. It's the good luck charm his French girlfriend, Jeanette, gave him back in Lorient in 1942.

U-505 sailor Werner Reh sees it happen to Goebeler and fears he'll lose his wristwatch.

"Put it in your mouth so they won't find it," a German whispers to him.

Too late. An American sailor swipes the watch off Reh's wrist.

Paul Meyer, second in command of *U-505*, is found floating on his life preserver with blood streaming down his face. He and a junior *U-505* seaman, Heinrich Brall, are transported to the *Jenks*. Meyer is taken to sick bay and treated by a pharmacist's mate. The nasty cut on his scalp is cleaned with tincture of green soap and Merthiolate, then sprinkled with sulfur powder and stitched with two sutures. Twenty-five pieces of shrapnel are extracted from his body. Meyer is given a haircut, shave, warm bath, and clean clothes supplied by the American Red Cross, then permitted to rest. He's offered a meal but says he has no appetite. Instead, he gulps down a cup of coffee. If the *kapitänleutnant* dies, Meyer will be the man blamed for this disaster. No wonder he has no appetite.

■ ■ ■

All this time, Gallery has been impatiently awaiting intelligence reports on the identity of the submarine and the name of her skipper. Finally, as the German sailors are plucked out of the ocean and taken on board *Chatelain* and interrogated, information trickles in.

"The captain is shot up and can't talk," Lieutenant Commander Dudley Knox radios. "Prisoners refuse to talk. Only information is 'Langer.'" It's a mangled version of Lange.

Gallery wants to make sure he hears right.

Knox repeats, this time with a slightly different pronunciation. "Prisoners say captain's name is 'Lenger.' No number on boat."

Five minutes later comes a clarification. "Captain of U-boat is Lange."

"Roger," Gallery says. "How long out of port?"

Knox: "We are trying to get that information now. Will let you know."

Two minutes later comes the skipper's confirmed identity. "Captain's name is Lange." But the POWs are already up to disinformation. "They claim the U-boat has no number. That it used the captain's name."

"Roger," Gallery says. He knows that's total B.S. Every U-boat is designated by a number.

This just in from Knox: "Latest information from survivors is they have been out three months or more. Nothing more could be gotten from them. Over." Then, an update: "Executive officer, or second in command, is named Lieutenant Meyer. Home is in Danzig."

"Roger," Gallery answers. "Report number of prisoners."

"We have fifty prisoners."

Next comes the confirmed identification of the boat.

"Twenty or more publications indicate the sub is *U-505*."

"Roger."

So, the submarine Gallery seized does have a name after all. She is called *U-505*.

Gallery ponders what to do next. It's clear to him that it's time to bring in Commander Earl Trosino. He can be a pain in the ass, but Gallery knows there's nobody on *Guadalcanal* who is more seasoned than his chief engineering officer in an emergency like this.

Trosino is in the engine room when he hears his name called over the aircraft carrier's loudspeaker.

"Commander Trosino, report to the captain on the bridge immediately."

Trosino sighs as he wipes his hands clean on a rag. "What the hell did I do now that that Irishman didn't like," he mutters.

He finds Gallery on the bridge. "Yes, sir."

"I want that ship," Gallery says.

Of course, Gallery is referring to the U-boat. He explains that he wants Trosino to board the submarine and save her from sinking and then tow the damn vessel to a US naval base. Trosino looks over Gallery's shoulder. The submarine's aft is submerged, and the fore is poking out of the water like the snout of a giant fish gasping for breath. What a clusterfuck.

"Sir," Trosino says, "I've never been aboard a submarine in my life." Then he adds, "I'm no hero."

"Never mind that. Pick the men that you want."

You never say no to a captain. Trosino salutes Gallery and goes off to find eight sailors who are about to "volunteer" for the most hazardous mission of their lives.

Trosino detests Nazis and their U-boat policy of targeting defenseless merchant ships. For him, it's personal. Back in 1942, he was an engineering officer on the USS *Long Island* when a communications officer woke him up at 4:00 a.m., never an hour for good news.

"Earl, what was the name of that ship your brother is on? The *Atlantic Sun?*"

"Yeah." Al Trosino was the engineering officer on the merchant tanker, which had departed Beaumont, Texas, on March 16, 1942, carrying 156 barrels of crude oil, en route to Marcus Hook, Pennsylvania.

"She just took two torpedoes a few minutes ago."

Trosino shook himself awake. Was this a dream?

"What did you say?"

"The *Atlantic Sun* took two torpedoes."

"Jesus. Keep me informed."

For the next two weeks, Trosino lived in a state of anxiety. Information about the *Atlantic Sun* was sketchy, and no one could tell him about the condition of the tanker or the casualties suffered. He had no idea if his brother Al was alive or not.

Trosino was granted shore leave and took a train back to Chester, Pennsylvania. His heart was hammering as he walked into the house and calmly asked his mother, "Did you hear anything from Al?"

"Yeah," Mama Trosino answered. "He's upstairs asleep."

Earl ran up the stairs and found Al sleeping. He slapped him across the face. Al bolted up in bed.

"What did you do that for?"

"For the worry I had for the last two weeks."

So, as a former merchant marine himself and the brother of a merchant marine whose ship had been torpedoed, Earl Trosino has good reason to abhor the Nazi policy of unrestricted submarine warfare. Trosino hates Nazis, and he really hates submarines.

■ ■ ■

Next, it's Lieutenant Deward "Hap" Hampton's turn to be summoned to the bridge.

"Lieutenant Hampton, report to Captain Gallery on the bridge."

Hap Hampton had been commissioned an ensign in June 1941 after graduating from the University of North Alabama. He has served on *Guadalcanal* since its shakedown cruise in 1943. Four weeks ago, when *Guadalcanal* sailed out of Norfolk on Mother's Day, Gallery ordered him to handpick a boarding party, just like Lieutenant David's on the *Pillsbury*. Hampton tried to envision what fields of expertise he'd need. He settled on a chief photographer's mate, whose job would be to take photos of the U-boat's equipment in the event she can't be kept afloat. Also, a crack electrician's mate, machinist's mate, chief carpenter's mate, engineer, boatswain, a signalman, and a yeoman to collect all of the U-boat's paperwork and documents. Hampton did his best to find a sailor who spoke German, but no such luck. As they crossed the Atlantic, the boarding party trained every day, never dreaming Captain Gallery's cockamamie plan to seize a U-boat would actually pan out.

When he gets to the bridge, Hampton salutes Gallery.

Gallery points to *U-505*.

Hampton can barely see it in the distance because the boat is almost submerged.

"That's your command over there. Good luck."

Say what? Hampton, who is twenty-five, plans on making the Navy his career, and it is his ambition to be named skipper of a great surface vessel one day. But it never dawns on him that his first command would be a Nazi U-boat. Maybe now is not the greatest time for the young lieutenant to inform Gallery about a very essential fact—he's claustrophobic. His fear of enclosed spaces never became an issue on the *Guadalcanal* because Hampton's job was mostly confined to the bridge and in the fresh air.

Hampton is ordered to link up with Earl Trosino's boarding party on *U-505* and offer technical and material support.

Hampton gives Gallery another smart salute. Apparently there are going to be two boarding parties from the *Guadalcanal*. Trosino is leading one and Hampton will command the other.

As Hampton heads to a whaleboat on the port side of the hangar deck, he can see Trosino and his hastily assembled boarding party lowering their whaleboat into the ocean. Next stop: *U-505*. That was quick. How the heck did Trosino pull that off? Hampton's head is swirling. *What am I doing here? I'm from a barely there farm in Alabama. I've never been on a submarine in my life.*

Trosino doesn't like this mission one bit. Wouldn't you know it, it's about to get worse. As they approach *U-505*, a great swell lifts Trosino's whaleboat out of the ocean.

"Pull forward! Pull forward!" Trosino yells at the motorman.

Too late. The whaleboat lands on top of the submarine with a shattering crash and a thousand pieces of splintered wood. Trosino pops out of the wave like a porpoise and he and his men are sent sprawling across the top deck.

"Grab the gear!" Trosino calls out. Christ, there goes the handy billy. He had intended to use the portable pump to drain the flooded sub, and now it's washed overboard, along with every other piece of equipment he had brought with him: five pairs of asbestos gloves, two sledgehammers, two fire axes, two hand saws, thirty pounds of nails, two monkey wrenches, four fire extinguishers, seven chisels, and two screwdrivers. Now what's Trosino going to do?

Chief Petty Officer Raymond Jackson leaps off the wreckage with a first aid kit slung over his shoulder. The chief photographer's mate, Clifford Whirler, manages to save all his camera gear. All well and good, but they can't salvage a sinking submarine with a first aid kit and a camera.

Trosino and his men find themselves marooned on the deck of *U-505*. Trosino is drenched. He doesn't know how Jackson and Whirler managed it, but somehow, they are bone dry—their shoes aren't even wet.

Next big problem: The wave has deposited them on the bow of *U-505*. The conning tower is way down in the middle of the boat. Everything between them is underwater. How the hell is he going to make his away across?

Trosino assesses the situation. Then he sees a guy-wire, tensioned across the length of the sub, from the bow to the conning tower. To test its strength, Trosino hangs his full weight on the wire. It doesn't snap. In a pinch, this might work.

Trosino proceeds to traverse the guy-wire, going hand over hand, without gloves. Along the way, he's battered by relentless waves. Signalman Gordon "Flags" Hohne from the *Pillsbury* boarding party is watching Trosino's progress from his perch on the conning tower. By the time Trosino reaches Hohne, he's so "doggone weak" he doesn't have the physical strength to make it the entire way.

Hohne reaches down and grabs Trosino by his hand and pulls him up.

When Trosino catches his breath, he asks Hohne, "Who's in charge?"

Just then, a head pops out of the hatch. It's Lieutenant Albert David.

"He's in charge," Hohne says.

"I'm Commander Trosino from the *Guadalcanal*, sent over to take charge of all salvage operations."

David gives a curt nod.

"What's it like down below?" Trosino asks. "Everything secure?"

"We got things going. There's nobody down below."

"Well, let's go below."

David ducks back into the hatch and Trosino proceeds to follow him. But first he has to step over what he thinks is a US sailor taking a nap on the deck. The guy is just lying there, his heels tilted up. What a knucklehead. Trosino kicks at his feet, not too gently either, and says, "Pardon me."

The guy still isn't budging. "Come on, bud, get up!" Trosino snaps. "Out of here!"

Now Trosino catches a glimpse of the man's face—mouth agape and tongue hanging out. His chest has been blown out by gunfire. It's the German radio operator Gottfried "Gogi" Fischer.

Oh, Jesus. Trosino turns to Hohne.

"Send word to the carrier: permission to deep six and bury this cadaver here."

Word comes back from *Guadalcanal*. Permission to bury the German sailor granted.

No time for ceremony. They roll Fischer off the submarine and the body floats away.

Trosino joins David down below.

The first sensation that wallops Trosino like a punch to the gut is the smell. Lord have mercy. Taking in that whiff, Trosino expects to find the submarine infested with "sixty million fleas." What a horrible and rancid stench. He tries to break down the components. Diesel fuel. Human body odor. Also, dank salt water floating with debris. Mildewed wet clothes.

How in the world can men live in this environment for months on end?

Trosino is startled to find himself standing in six inches of water covering his shoes. He ignores the soaking and looks around.

From his knowledge of German ships, Trosino knows there's got to be a well-stocked set of tools lying around. A wrench of some sort. Any tool. Sure enough, he finds a T-wrench stored on the side of the periscope. Good. Now he's got something to work with. A few turns with a T-wrench can do wonders.

With Lieutenant David at his side, Trosino inspects the submarine. What an intimidating environment. No simple steel cigar, it's a labyrinth of components—"a great big, huge tube of nothing but valves and machinery." In the engine room he wades through two feet of slush—a foul muck of water and oil. The bilge below the control room is filled to the brim. With each passing minute, U-505 is sinking farther down by the stern. Where to even begin?

"We've got to stop this thing from spinning so we can do some work," Trosino tells David. "Get everybody up on topside. Get 'em out of here 'cause I'm gonna stop her." Smart move, considering what happened last time.

Trosino and David order the boarding party out. The last man out closes the conning tower hatch behind him.

There are only two men left inside U-505: Trosino and David. They proceed to close all the fore and aft watertight compartments. When they're finished, Trosino tells David that he needs to rejoin his men.

"I'm not going topside," David replies.

Trosino looks at him. "Lieutenant, this is a one-man job, and I'm not asking for volunteers. I'm the senior officer here in charge. You get topside."

David grins at the commander. "You're the kind of man I like to work for." Then he insists, "I'm staying."

Trosino gives up. It's David's funeral. "Well, I'm appreciative. Misery loves company."

Trosino and David find themselves standing before an electrical circuit board. One wrong move, and they risk energizing all the demolition charges. David's glasses are fogging up. He keeps wiping them. Beads of cold sweat speckle his forehead.

"Lieutenant, how do you feel?"

"Oh, I feel all right. How do you feel, Commander?"

"I'm not gonna lie about it, I'm damned scared. What I wouldn't give to be under an apple tree any place in this world other than right here."

Time to do or die. Trosino makes an educated guess and pulls a lever. That does it. The engines stop, and so does the submarine. The steady electrical hum suddenly goes silent. The propeller is dead in the water. Unfortunately, once again, U-505 is dipping down due to the loss of forward momentum. Definitely it's time to get out of there.

They climb up the control room ladder, David first, Trosino right behind him. For all they know, opening the hatch will send the ocean flooding into the ship.

Trosino tells David, "If we're underwater, you go out first and then I will follow you. You go up with a bubble of air."

Sure enough, David opens the hatch and a gush of green water knocks him right back. David hastily closes the hatch. There isn't much else for them to do other than try again. They certainly can't remain in the sub. Deep breath this time. David shakes Trosino's hand to say goodbye.

Trosino isn't so sure the end is near. "Look, you get up there and get close. Open up and I'll push you out." Sounds like a plan.

David climbs the ladder. Right behind him, Trosino puts his hands on the lieutenant's butt to give him that extra nudge.

"Sing out when you're ready," Trosino says.

"Okay, Commander."

David opens the hatch, but instead of ocean they see glorious sunlight. What luck—U-505 must be riding a hollow trough between two wave crests. David takes another step up the ladder and finds himself on the conning tower bridge. Trosino follows. They secure the hatch.

Neither man can say anything. Trosino tries to speak but he can't. He and David just stand there, shaking. For what they've just lived through, there are no words.

Everyone is topside, including all nine sailors from Lieutenant David's boarding party. They take a moment to collect themselves. It's beginning to dawn on them that they've just pulled off the impossible.

But Zeke Lukosius doesn't feel as if he's home free. On the way up the ladder, he stumbled into a spiky projectile and sliced his Mae West life jacket, a big deal because he really doesn't know how to swim. All he can do now is hope the submarine stays afloat and he doesn't slip overboard.

And Chester Mocarski is in bad shape. His body is lying on the deck, half in water, half out. He's moaning in agony and hasn't moved since his legs were crushed when he got slammed between the submarine and the whaleboat. David turns him over and Mocarski's deep blue eyes stare back blankly. They carry him into the whaleboat and make him as comfortable as possible under the boat's awning, but the pain is unbearable.

Coxswain Philip Trusheim gives him an injection of morphine. Then Jim Beaver, who did such an outstanding job snagging the whaleboat to the submarine with that bow hook, takes a felt pen and marks Mocarski's forehead with a bold letter *M*. That's so the medics who treat him later will know Mocarski has already been administered morphine. Any more could result in an overdose.

Mack Pickels is generous with his booty. He hands Trusheim a pair of binoculars. He also gives a pair to Beaver, and another pair of binoculars goes to the third member of the whaleboat crew, Motor Machinist Robert Rosco Jenkins, who comes from the humble town of Gypsy, West Virginia, population on a good day maybe one hundred country folks.

There's no way to exaggerate the magnitude of their accomplishment. They have in their possession the Nazi code books, charts, submarine operating instructions, and an Enigma machine. It's certainly one of the great intelligence windfalls of the war. Then, of course, there's the submarine itself, intact and loaded with the latest German technology, including acoustical torpedoes. Assuming, of course, they can keep her afloat.

Naturally, the cigarettes come out. They all light up. David's hands are trembling. He's shaken and knows he has no business being alive. The same goes for all of them.

He turns to Radioman Third Class Stanley "Stash" Wdowiak, the kid from Brooklyn who, back on the *Pillsbury*, expressed skepticism about the plot to seize a Nazi U-boat. David remembers that conversation after an afternoon of training, when Stash asked him, "Do you think the Germans are just going to let us come aboard? Huh?"

David takes a drag on his cigarette. "So, you don't think the Jerries will invite us on board, huh, Stan?"

Wdowiak can only shake his head in utter disbelief. Still jittery from the experience, he says simply: "I don't know, Mr. David. I just ain't talking."

TWENTY-SEVEN

REQUEST IMMEDIATE

ASSISTANCE

As usual it's Earl Trosino bringing the dreamer back to reality. Trosino signals to Captain Gallery that if *U-505* is to be saved from a fate on the ocean floor, she's going to need a tow right away.

Gallery sends the USS *Pillsbury*.

As *Pillsbury* steers for *U-505*, Trosino realizes he's watching a slow-motion catastrophe in the making. The *Pillsbury* should be approaching *U-505* by her stern, while throwing the towline to the submarine's bow. In other words, the back of the *Pillsbury* to the front of the submarine. Ass to nose, so to speak. Instead, the ship is coming alongside.

Ocean towing is a complex operation under the best of conditions, but towing *U-505* is especially challenging because the sub is half-submerged and still taking on water. Also, like an iceberg, maybe 90 percent of the submarine is below the waterline.

Sure enough, as the USS *Pillsbury* draws near—collision! *U-505*'s diving plane, which sticks out of the submarine like a flipper, slices a deep gash into the *Pillsbury*'s thin steel skin.

Trosino shakes his head. Lieutenant David also can't believe it. Are they seeing things? After all they've pulled off, *this* is how it goes wrong?

Now George Casselman on the *Pillsbury* has the difficult task of letting Captain Gallery know what just happened. As the *Pillsbury* limps away, he radios Gallery via Talk Between Ships.

"We backed away from the sub now. Her planes made a big hole in the side." He reports that two main compartments on the *Pillsbury*, including the forward starboard engine room, are flooded with four feet of water. The *Pillsbury* is at risk of sinking—not by a U-boat torpedo but by ineptitude.

"Are you making any progress on sub?" Gallery asks Casselman. He's still totally focused on *U-505*.

"Not much. Over."

Gallery is informed that the submarine's rudder is jammed and power has been cut. The sub is dead in the water. And she's "settling slowly."

Meaning—she's sinking.

Casselman tells Gallery that in his opinion the sub can't be saved "unless they get a tow." Which echoes what Trosino has been saying.

"We tried it," says a chastened Casselman. "The [diving] planes made a job for us. Whether anyone else could do it, I don't know."

Gallery has heard enough. He's not about to lose *U-505* now, not when he's this close.

"I am going alongside myself to put towline on myself," he radios. Then he roars, "Keep out of my way."

As *Guadalcanal* bears down on *U-505*, Gallery gets his own up-close assessment. It's more dire than he anticipated. The sub had been listing about 10 degrees by the stern, but now that *U-505* is dead in the water, the tilt is even more extreme. Waves are breaking over the conning tower hatch. There's no way of getting inside the boat now.

Gallery eyeballs Trosino. There's Albert David and all the sailors from the *Pillsbury* and the *Guadalcanal* boarding parties. He sees Lieutenant Hap Hampton's boarding party lined up on the sub's bow, twenty feet of which is jutting out of the water. Only three feet of the conning tower remains above water. The sub's stern is completely submerged. *Jeez.*

Gallery doesn't have much time. He steers the aircraft carrier as close as he dares get to the submarine's "ugly snout." He's practically "nuzzling" *U-505*'s front torpedo tubes. At this point he's worried some dimwit will accidentally push the wrong button and launch a torpedo straight into the aircraft carrier. Gallery says a silent prayer. "Dear Lord," he mutters, "please don't let any of them monkey with the firing switch." Another notion pops into his head: *Suppose there is just one Nazi left on that sub . . .*

An inch-and-a-quarter-thick steel towing line is fired to Hampton's crew on *U-505*. Heaving and slipping on the wet deck, about a dozen sailors give it their all and drag the line to the bow of the submarine. With some fiddling and tinkering and a lot of muscle power, they rig up a bridle. Fingers crossed, that should hold.

Guadalcanal eases ahead. The towline is stretched and as Gallery gathers speed, *U-505*'s stern emerges from the water until the tow is as "taut as a banjo string." The sub achieves buoyancy. A great cheer arises from the men, and Gallery breathes a sigh of relief.

Gallery figures he can now put the stamp of victory on his historical achievement. Nothing will symbolize that more than the American flag. Old Glory is hoisted above the conning tower. Now *U-505* can officially be said to be the first enemy ship to be seized in battle since the War of 1812.

Earl Trosino heads down the hatch of *U-505* once more and makes his way to the aft torpedo room. He wants to check out those German acoustic torpedoes, but he finds the watertight door jammed shut. Trosino starts slamming away with a sledgehammer. Then a crowbar. For the life of him, he can't pry the door open. But maybe that's a good thing because he has no clue what awaits him behind that door. What if the torpedo room is flooded? He could unwittingly unleash a deluge and sink the sub. Reconsidering, he lays down the sledgehammer and crowbar.

As Trosino makes his way through *U-505*, he doesn't like what he's seeing. Several sailors from Hampton's boarding party are scurrying about the submarine, and they seem intent on hunting for souvenirs. Some of the guys are stuffing German time bombs into their back pockets! Seriously? Handled the wrong way, even disconnected, those devices still contain explosives and could explode.

"Get these men out of here!" Trosino snaps. He orders a gunnery officer to collect the bombs and throw them overboard.

Trosino sees a young sailor parking his butt in the German skipper's chair. Another goofball is running the periscope up and down just for the heck of it. They're like kids who have broken into a toy store.

"Get 'em off the ship," Trosino growls.

Next comes an important job—what to do with the shit bucket. The stench is making everyone nauseous. Trosino asks one sailor, then another and another, to help, not as an order, more like a favor. Nobody takes him up on it, so Trosino performs the stomach-churning job himself. The commander proceeds to roll the shit bucket out of the engine room and then he dumps the contents straight into the bilge. Of course, a photographer's mate is there to record Trosino's derring-do for US Navy historical posterity. Trosino is wearing a straw Panama hat that he found on board *U-505* to replace the regulation Navy hat that got swept into the Atlantic when he crash-landed on *U-505*. In the straw hat he looks like a humble peasant, performing a task nobody else would stoop to carry out.

Trosino supervises the salvaging operation until darkness falls. Gas-powered pumps positioned on the deck of *U-505* are finally draining the water out of the sub. The immediate danger of sinking, for now, seems to be receding.

That night, a whaleboat transports Trosino back to the *Guadalcanal*.

Wet and exhausted by the day's drama and physical exertions, Trosino has one more duty to perform before catching some shut-eye. He reports to the bridge and informs Captain Gallery that *U-505* has been battened down for the night and that it is his professional opinion the sub will stay afloat at least until morning. But he expresses concern about the aft torpedo room. He can't be certain it isn't flooded, because the watertight door is wedged tight and he didn't want to force it open for fear of unleashing a cascade of water. It also looks to him as if the door might be booby trapped.

Gallery takes it all in. Trosino has had a horrendous day, crawling through oily water, squirming under the engines, and even disposing of human excrement in a virtuoso endeavor at keeping *U-505* afloat. Now Gallery must let his chief engineer know that he's to get a good night's sleep and then return to the submarine in the morning to fix the rudder, which is still jammed hard right and happens to be situated in the aft torpedo room. One way or another, they've got to get that damn door open.

In other words: Rest up, Trosino. We need you strong to risk your life again tomorrow.

Then Gallery says something he's never said before. He calls Trosino by his first name.

Lieutenant David may have captured the sub, but it's Trosino who saved it.

Executive Officer Jessie Johnson puts in his two cents.

"You'll get a medal for this," he tells Trosino.

The weary commander isn't impressed. "Medals don't buy pork chops," Trosino answers. But then he adds, "I'd be proud to wear a medal if everybody in the task force is recognized."

■ ■ ■

The sun is setting when the whaleboat returns Albert David and the boarding party to the *Pillsbury*. Lieutenant Commander Casselman is there to greet his men. He's got a bottle of Canadian Club in hand to celebrate.

Flags Hohne is having a delayed reaction to the day's events. He now sees there's a hole in his life jacket. Had he gone over the side, he could have ended up fish meat. Hohne's body starts shaking as the full realization of his experience on *U-505* finally hits home. He utters a little prayer and thinks, *How lucky am I to be alive.*

"How'd you like a drink?" Casselman asks.

Hohne gratefully accepts a shot of whiskey to steady his nerves.

Gunner's Mate 1st Class Chester Mocarski is taken to sick bay, the letter *M* still marking his forehead. An examination shows no evidence of shock or obvious bleeding. All his injuries appear to be internal. The doctor prescribes plasma followed by morphine every four hours.

There's another patient in sick bay—Kapitänleutnant Harald Lange. *Pillsbury's* medical staff is doing what it can to save the right leg of *U-505's* skipper, but it's looking hopeless. Lange's leg will probably have to be amputated.

Pillsbury is in a state of crisis. Flooding in the engine room has knocked out the destroyer escort's power and the ship is crippled.

No rest for Mack Pickels. He's told his presence is needed below deck, and he heads straight for the engine room. He finds it in shambles. Mattresses have been piled against the five-inch-wide, twenty-inch-long gash in the *Pillsbury's* hull to minimize flooding. Carpenter's mates have shored up the leak with lumber. They've been waiting for Pickels to assist in rigging the collision mat, which is a large canvas square treated with a sealing agent and tied down with thick rope on all corners. It functions as a temporary patch and fits over the gash like a giant bandage.

Submersible pumps clear out the water, and by 11:00 p.m. the engines kick in and power is finally restored. The lights go back on.

Time for some chow. Pickels is starving. He hasn't eaten since breakfast. Man, oh man, what a day. Pickels heads for the mess hall. And wouldn't you know it, they're out of grub. All the cooks can drum up is a bologna sandwich. Terrific. Apparently, those hungry German POWs taken on board have gobbled up all the leftovers.

Back on *Guadalcanal,* the time has come for Gallery to let his bosses at Atlantic Fleet Headquarters and the commander in chief of the United States Fleet know about the day's historic accomplishment. Gallery sends the following message, which he is confident will be forever enshrined in naval lore. He keeps it short:

REQUEST IMMEDIATE ASSISTANCE TO TOW CAPTURED SUBMA-RINE *U-505.*

Gallery chuckles. The duty officers back in Norfolk and the Pentagon will be in a tizzy once they get that message decoded.

The response comes quickly. There must be some mistake, some error in the decode. Recheck and repeat message.

Gallery suppresses a smile. He sends confirmation:

HAVE *U-505* IN TOW. KAPTAN LIEUTENANT LANGE. 80 DAYS OUT.
49 [SIC] PRISONERS. DON'T HAVE FUEL FOR CASABLANCA.
REQUEST PERMISSION PROCEED DAKAR.

Dakar, Senegal, in what is then known as French West Africa, is about
four hundred miles away. It is the nearest friendly port, but it won't be
an easy voyage. Towing *U-505* is a drag on *Guadalcanal*'s speed—she can
only make six knots. The aircraft carrier's maneuverability is also limited
by the towline. The *Guadalcanal* is dead meat if any U-boat can suss out
the situation and launch an attack.

Gallery shuffles the task force. The USS *Pope* is dispatched to stand by
the stricken *Pillsbury*. The three remaining destroyer escorts—*Flaherty*,
Chatelain, and *Jenks*—are ordered to form a protective screen around the
Guadalcanal in the event she comes under attack by a German wolf pack.
Gallery also sends two fighters into the air to circle the task force and
look for Nazi subs.

But what's taking so long for the bosses to get back to Gallery? He
radios another coded message reiterating that he's setting course for
Dakar.

That wakes up Atlantic Fleet. It's a prompt rejection. Gallery is told to
sit tight.

GUADALCANAL GROUP WILL NOT REPEAT NOT PROCEED DAKAR.
FURTHER ORDERS COMING SOON.

Dakar is a beehive of intrigue and Nazi spies. Towing a captured Nazi
submarine into port would have gotten back to Berlin like a news bulletin. The Germans even have an embassy there.

■ ■ ■

A lot happens overnight.

Six minutes before midnight, the towline attached to *U-505* snaps.
Gallery isn't surprised. An aircraft carrier isn't built for towing. He probably put too much stress on the line, because he's zigzagging as a defense

against enemy attacks. There's nothing to do but wait until morning for light, at which point he'll toss over another line—this time the thicker two-and-a-quarter-inch wire. Just in case the situation deteriorates further, the carrier stays close to the sub, circling it for the rest of the night.

At last Gallery hears from Admiral Royal Ingersoll, commander in chief, Atlantic Fleet. Gallery is ordered to change course immediately for the British colony of Bermuda. And assistance is on the way. The tugboat *Abnaki* has been split off from a convoy bound for Africa and dispatched to take over the towing of *U-505*. Also, the oil tanker *Kennebec* will rendezvous with *Guadalcanal* and the destroyer escorts, and refuel Task Group 22.3 in the mid-Atlantic.

Gallery shakes his head. *Bermuda*. Whose idea is that? Chrissakes, that's 2,500 miles away. A fifteen-day tow clear across the Atlantic. That would make it the longest tow ever. Nobody has ever tried to pull off anything like that.

■ ■ ■

Dawn, June 5, 1944.

A big day ahead. Top priority is attaching a new towline to *U-505*.

Before he gets going, Gallery sends a communiqué to Casselman over on the USS *Pillsbury*. He knows Casselman must have tossed and turned all night, sick with worry over that collision with *U-505* and the impact to his naval career. Casselman could face discipline for dereliction of duty once they return to America.

He's worried to death, until he gets a message from Dan Gallery.

THIS IS FOR YOUR FILES REGARDING DAMAGE DONE TO YOUR SHIP THIS CRUISE. THIS DAMAGE WAS DONE EXECUTING MY ORDERS AND I ASSUME RESPONSIBILITY.

A class act.

TWENTY-EIGHT

JUNIOR

Forty-seven German prisoners spend their first night in American captivity on board the destroyer escort *Chatelain*, while the remaining POWs are scattered between the *Pillsbury* and the *Jenks*.

And now the question is: Who's gonna talk?

On the *Jenks*, Paul Meyer, second in command of *U-505*, is undergoing interrogation by a young gunnery officer, Lieutenant Frederick Slack, a Dartmouth graduate from Philadelphia. Slack's father is president of a coal company and belongs to the prestigious Union League and a swanky Main Line country club. But the main thing about Slack is that he once lived in Germany and is fluent in the language. If anybody can get through to Meyer, it's Frederick Slack. He gives it a shot, but Meyer doesn't give up much.

They have better luck squeezing *U-505* prisoner Heinrich Brall. The seaman is only nineteen and scared out of his mind. Someone on the *Jenks* has the inspired idea of keeping Brall isolated from Meyer, so he's locked in the ice machine room where the ship's perishable food is stored. Brall is guarded by a twenty-one-year-old machinist's mate second class, Carl Tuchalski, from Staunton, Illinois. Tuchalski is fluent in German and laid back in his interrogation technique. As he questions Brall, he's also tinkering with the ice-making machinery. It seems to relax Brall, and they start engaging in an easy exchange—two working stiffs who are good with their hands, about the same age, who happen to be fighting for different sides in the world conflict.

Before long, Brall is yammering away. He tells Tuchalski all about himself: that his birthday is next month, and that he worked on the docks for two years before enlisting in the Kriegsmarine submarine corps and that this was his first war cruise. Brall is also chatty about *U-505*—that she carried fourteen torpedoes, sixty men, four officers, and had been at sea about eighty days. He says the boat had sighted three enemy vessels during this cruise, but hadn't sunk any.

Tuchalski duly reports all this intelligence to his commanding officer.

On board the *Chatelain*, the thirty-eight German POWs are issued clean clothes and toothpaste and toothbrushes. To relax and make them feel at home, a Viennese waltz is played over the loudspeaker. Joe Villanella, the radarman from Newark, New Jersey, can't quite understand the cordiality. *We're offering these Germans entertainment? This is nuts!* On his own initiative, Villanella shuts off the music.

Then he sees a POW wandering about. What the hell is he doing up here? He should be in the brig with the other prisoners.

Villanella steps out of the radio shack. Turns out, the prisoner speaks a little English, enough to get by.

"We shot a torpedo at the carrier but I think your depth charges must have knocked it off course." Spoken oh-so-casually, and as if he regrets the torpedo missing.

Villanella gets in the German's face. "Do you think you'd be standing here talking to me if you'd sunk that carrier?"

The German looks quizzically at Villanella. "You would kill us?"

"I would kill *all* of you in the water. You wouldn't be on this ship. If you'd killed a thousand guys on my ship, you can expect the same from me."

The German sneers at what he considers unprofessional military behavior. "You are not Navy," he declares, and walks away.

Under heavy guard, with Tommy guns pointed at them, all fifty-nine POWs from the *Pillsbury*, *Chatelain*, and *Jenks* are transported by whaleboats to the *Guadalcanal*. Up the Jacob's ladder they climb, while Harald Lange is hoisted aboard on a stretcher and whisked to sick bay.

Everyone else is herded to a catwalk below the flight deck for showers. A high-powered hose sprays them with salt water drawn directly from the ocean. Not what you'd call a gentle mist, but it's a relief to wash the diesel fuel off their bodies. They're strip-searched one more time, including a cavity probe. Petty Officer Karl Springer is startled when his ass cheeks are squeezed. It's not a humiliation tactic. The story going around is that Japanese POWs have been known to conceal hand grenades between their buttocks.

Only the officers are taken to the brig while the rest of the POWs are locked in a caged pen located below the flight deck. The noise of planes taking off and landing is earsplitting. Sleep in this racket? Not a chance.

In no time, they are all drenched in sweat—hot air is blasting right at them from the carrier's ventilation system. It's a sauna in there. One guard takes pity and shifts a large standing fan over to cool down the prisoners, but when his superior sees this, he moves the fan back. No mercy must be shown these Germans.

Gallery takes a few minutes to inspect the conditions and shows up just as a guard is walking among the prisoners, inside the pen—carrying a Tommy gun. Jesus Christ! Who does he think he is—Wyatt Earp? He's outnumbered forty to one. What's to stop the Germans from overpowering him, seizing the Tommy gun, and even taking over the aircraft carrier? Improbable? Maybe so. But just as improbable as seizing a U-boat on the open sea. Gallery throws a fit and chews out the guard, telling him he doesn't want to ever see anything like that again. Think, man, *think*! Gallery has no intention of mistreating the prisoners, but that doesn't mean he has to trust them or get chummy. These U-boat characters are a hard bunch of "desperate men."

The realization that *U-505* has fallen into enemy hands is making the POWs despondent. And so is the mockery from the American sailors, which is really getting to them. At first, the Germans don't believe these gum-chewing specimens of a decadent democracy have outwitted them. Grinning buffoons. An Ami trick—Ami being German military slang for "American," or, in German, Amerikaneri. *U-505* seized? Surely, it's nothing but a ploy to get them to spill U-boat secrets. But every now and then, *Guadalcanal* rises on an ocean swell, and that's when the POWs catch sight of their beloved *U-505* and realize she's being towed by *Guadalcanal*. Most distressing of all, she's flying an American flag.

So, it's undeniably true. *U-505* has fallen into enemy hands. They have failed in their most sacred duty to scuttle or go down with the ship.

Nobody takes it harder than the last man out, Hans Goebeler. Where did he go wrong? He tries to replay the last moments in his head. He thought he had thrown the sea strainer cover straight into the bilge. But maybe not. Maybe he had carelessly tossed the cover aside, never fathoming it would be discovered by the American raiders.

At least the chow is decent. The Germans are fed what the US sailors eat. They are also taken out for calisthenics, bending and stretching, but they are not permitted to shave or get a haircut, even if their beards are itching like crazy in the nasty heat. Rumor has it the Americans want them to look like vagabonds for propaganda photos when they are taken ashore. The bellyaching is nonstop. The Germans complain about everything, including the lukewarm drinking water.

Hans Goebeler barely notices the inconveniences. He's obsessed by a single thought, playing over and over in his head: *This is all my fault.*

■ ■ ■

Summoned to the bridge, Lieutenant Jack Dumford finds himself in the intimidating presence of the great man himself, Captain Daniel Gallery.

And he can't help wondering: *Why am I here?*

The twenty-five-year-old communications officer from Belleview, Kentucky, hasn't had much contact with Gallery in all the months he's served on the *Guadalcanal*. Mostly, his duties consist of decoding messages that come over the radio, then handing them on a clipboard to the communications watch officer, who then initials and disseminates them to the XO before the message is turned over to the captain.

But now all those by-the-book steps that have kept Dumford and Gallery apart are gone. Gallery is looking Dumford right in the eye as he tells the young officer he's to report to a whaleboat because he is next in line to board *U-505*. His mission is to scour the submarine with fresh eyes and retrieve all remaining code books, navigational charts, and sensitive documents—anything related to intelligence. Make sure the prior boarding parties haven't overlooked anything and haul all the top-secret material back to the *Guadalcanal* for analysis.

A mop-up mission, but a vital one.

Dumford salutes Gallery and reports to the whaleboat straightaway. He climbs on board, and as the tiny craft is lowered and slices its way across the ocean, he feels a knot tightening in his stomach. He doesn't mind admitting he is scared to death and wonders if any of the other sailors on the whaleboat remembered to bring toilet paper. Just kidding . . . but not really.

Even Dumford must admit he's not a natural born sailor. He'd been captain of his high school basketball team and a student at Morehead State University when he enlisted in the Navy. His family had connections in Washington, and the next thing he knew he was sworn in as an ensign.

Until now, his time on board the *Guadalcanal* has been enjoyable. The radio room pretty much ran itself. The radio technician, warrant officer, signalman, and communications officer were all career sailors who knew what they were doing. Dumford spent his days playing cards, drinking plenty of coffee, smoking too many cigarettes. A lot of leisure time took place in the wardroom with the other junior officers. He relaxed in the sun and tried to grow a beard. His only real physical discomfort came one night when he ate pork chops for dinner and got food poisoning. An old salt told him, "You shouldn't have had pork chops." Lesson learned.

But what is happening right now, this is the real deal. He's about to board a captured Nazi submarine.

Dumford doesn't have far to travel. The submarine is under tow right behind the *Guadalcanal*. Just the same, the waves make it rough going. Jumping onto the U-boat, Dumford is certain he's going to slip overboard and drown. But he manages to hang on and make his way to the conning tower. The deck is slippery and those hard-sole shoes he's wearing are not very practical. How he wishes he'd changed into his basketball sneakers.

Down the hatch goes a quaking Dumford. He's pretty certain he will not emerge alive. The submarine will sink. Or a booby trap will detonate. Either way, he accepts the fact that he'll never again see Lillian, his beautiful Kentucky bride, again.

He takes a deep gulp and finds himself inside *U-505*. If he's ever sent to hell, it'll look just like this, bleak and downtrodden, a nightmare. What a miserable existence those German sailors must have had! Nothing in the world would make Dumford serve on a submarine. As he passes the mess hall, he sees remnants of the Germans' breakneck evacuation—dirty dishes are everywhere.

Dumford locates the radio room and gets to work. His fascination overcomes his fear as he stuffs every relevant document and chart he finds into his seabag. There are "tons" of decoded messages. Into the bag they go. There's no time to make an inventory or examine the material, not that he could because he doesn't know a word of German. He just knows it's a gold mine of intelligence that, once analyzed, could expose the U-boat fleet's methods of operations. He fills nine canvas bags before making a hasty retreat.

Of course, there's always time for a souvenir. Dumford sees a Luger, which he tucks into his pocket. He also grabs a knife whose blade is engraved with the initials KB.

When Dumford returns to *Guadalcanal*, he's so worked up about what he's found that he heads straight for Captain Gallery's cabin. He wants to make sure Gallery knows what a big deal this is.

Gallery gives him the brush-off, as if to say he doesn't need a lieutenant junior grade telling him what he already knows.

Over the next few days, Dumford keeps at Gallery, until, in his words, he's getting to be a real pain in the ass.

Dumford gives it another shot. All he wants is a few minutes alone with Gallery. Dumford knocks on the captain's door, and he hears Gallery's voice telling him to come in. It's not a good time, Dumford sees right away. He's interrupting the captain's meal. Gallery sits there at his table slurping tomato soup. Over the rim of the bowl, Gallery's eyes narrow. *Not this guy again.*

"I know what's wrong with you," he tells Dumford. "You want to be a secret agent."

Dumford figures he's blown it with Gallery, but he's wrong. The captain will soon have a special mission for this wannabe spy.

■ ■ ■

Captain Gallery is keen to get on board *U-505* and check things out for himself. Earl Trosino offers him the perfect excuse when he tells Gallery that the torpedo room door may be rigged with explosive booby traps. It just so happens that Gallery did postgraduate work in ammunitions and weaponry and general naval ordnance. He knows all about fuses and circuitry and considers himself as qualified as anyone on the task group to deal with bombs. At least that's what he tells Trosino. It may even be true, but the fact is it's just the excuse Gallery is looking for. Right then, Gallery creates a new position: officer in charge of booby traps.

Gallery and Trosino take four sailors with them. Nothing on *U-505* has improved overnight. The sub is in worse shape. The bow is once again jutting out of the water and the stern is clear under. Ocean waves are breaking over the conning tower. Gallery's first reaction? He isn't certain he has "any business being there."

When they're all safely on the conning tower, they proceed down the hatch, which they have to quickly shut behind them to keep the waves from washing in. Dan Gallery looks around in wonder, knowing just how remarkable this moment is. He's given twenty-seven years of his life to the US Navy, and this is the first time he has ever stepped inside a submarine—an enemy sub, to boot.

Like everyone else, Gallery takes a whiff of that stink and makes a face. He also notices that the lights are very dim, which means the batteries are nearly drained. The vessel is also rolling and pitching and dipping by the stern.

Enough. They've got to straighten out this mess. Trosino leads the way, striding briskly past the control room and the diesel engines. The farther aft they go, the more Gallery is struck by the immense length of the sub. He's also seized by the same sense of fatalism that the other

American sailors felt when they stepped into *U-505* misery. Gallery handles it by telling himself a few simple words: "You can't live forever." Finally, they reach the torpedo room.

"There she is, Cap'n." Trosino shines a flashlight on the fuse box.

Gallery immediately sees the problem.

The fuse box has popped wide open and the cover is blocking the watertight torpedo room door. Inside the exposed fuse box is a tangled mess of interconnected and crisscrossing wires. The only way of getting into the torpedo room is to shut the fuse box. But in doing so they run the risk of energizing the circuitry.

"It may be the charge to a booby trap," Trosino tells Gallery. "I didn't want to close it."

Gallery has to agree the open fuse box is suspicious and has all the makings of a lure to ruin.

What now? He and Trosino ponder their next step. They gently probe and fiddle with the wires and don't discern anything dodgy. Gallery's coming around to the theory that the crew of *U-505* abandoned ship before they had a chance to set any detonation charges. What most likely happened is that the depth charges that exploded around *U-505* on June 4 shook all the fuse boxes open.

Of course, believing in a theory is one thing. Testing it is another.

"Here goes nothing," Gallery says. That phrase, or a variation of it, is becoming the unofficial motto of Task Group 22.3. The way Gallery figures, he'll know if he's right or wrong in about a half second. If he's wrong, he won't be around to berate himself for miscalculating—he'll be blown to bits, along with his chief engineering officer.

Trosino holds his breath, braces himself, and mutters a little prayer as his captain eases the fuse box cover shut.

It's all good.

Trosino can breathe again.

Time now to deal with the torpedo room door.

Gallery, Trosino, and one seaman brace their backs against the door and crack it open just a bit in case the torpedo room actually is flooded. If they see water come spurting around the edges of the door, they'll need to slam it shut right away. But not a drop of water comes through.

They open the door a little more, then all the way. It's pretty much bone dry in there.

They enter the torpedo room and look around.

"Wait a minute," Trosino says. He's pointing at the rudder indicator level. Left is port. Right is starboard. The little oil bubble is as far right as it can go, indicating the boat is totally unbalanced. Suspended below the vial is a mechanism that doesn't look at all like a wheel but is recognized by Trosino for what it is. "This is the manual steering," he says.

Gallery shoves Trosino aside. Then he yanks out a pin and brings the column down, and returns the pin, locking the manual steering in place. The entire process takes only thirty seconds.

"There," Gallery announces. "That'll hold. She'll tow."

They watch as the oil bubble on the rudder indicator floats back to the center, where it should be, amidship. The boat is now balanced.

"Let's get the hell out of here," Gallery says. No one objects.

They head back to the control room. It's uphill all the way. The passageway is still slanted at maybe a 15-degree gradient. Then they clamber up the ladder, and as they emerge out of the hatch they feel the delicious warmth of the sun on their faces.

And there's a big surprise, as well—seems that the boys on deck had gotten hold of a thick brush and a can of red paint to scrawl *Can Do Junior* in big bold letters across the conning tower. Gallery is delighted. *Can Do* is his nickname for the *Guadalcanal*. *U-505* now had a new name—*Junior*. He watches with pride as an American flag flies majestically on a jury-rigged flagpole atop the bridge of *U-505*. Below Old Glory is a smaller flag—the Nazi flag, with swastika. In Navy tradition, it is a symbol of victor over vanquished.

TWENTY-NINE

"I WILL BE PUNISHED"

They're doing whatever they can to save Kapitänleutnant Harald Lange's leg, but it's not looking good.

Lange tries propping himself in a sitting position. He is groggy, having just gone through a blood transfusion donated by one of his men. He can only see out of one eye due to the splinter that sliced into his eyelid.

He has a visitor. Captain Daniel V. Gallery introduces himself.

"Captain, my name is Gallery. I'm commanding officer of this ship."

Lange bows respectfully. Gallery reminds him of an old seafarer, like himself.

Gallery mentions his previous encounter with another German U-boat commander he recently took prisoner, Werner Henke.

Then Gallery drops the bombshell, informing Lange that *U-505* has been seized and "we have her in tow."

Lange is incredulous. Not possible, he retorts. Until this moment, he had assumed the crew had carried out his order to scuttle the boat, that *U-505* was sleeping with the fishes.

Gallery assures Lange that he is speaking the truth. Now it's time to bargain.

"Look here, boy. You are in a very bad situation with your leg," Gallery says. "I can send you in an aircraft to the best hospital in the USA within twenty hours. But in exchange you must help me a little bit now." Gallery leans in close. "What's the matter with the boat? Did you install anything which could make it explode?"

Lange shakes his head and tells Gallery he is too weak to talk.

The next time Gallery visits sick bay, he's got something to show Lange: a framed family photo of Lange's wife, Carla, daughter, Jutta, and son, Harald Junior. It's the picture Lieutenant David grabbed when he ransacked Lange's cabin. Only now does Lange realize that Gallery has been on the level. This is no Amerikaner trick. *U-505* has fallen into the hands of the enemy.

"I will be punished for this," he says with a sigh.

Gallery tries to soften the blow. He points out that Germany is losing the war. Soon, the Nazis will surrender, or be replaced.

Lange can't stop shaking his head. What difference does that make? Whatever government follows the Nazis, he will be held accountable. He will be court-martialed and disgraced.

"I will be punished," he repeats.

Humbled by the news of this fiasco, Lange starts cooperating with Gallery, prefacing most of his answers with, "I wouldn't tell you this ordinarily, but you know the answer anyway."

He agrees to make a one-page statement recounting how he came to be taken prisoner.

When I sat in the pipe boat [inflatable individual life raft] *I could see my boat for the last time. Some of my men were still aboard her, throwing more pipe boats into the water. I ordered the men around me to give three cheers for our sinking boat. After this, I was picked up by a destroyer where I received first aid treatment. Later, on this day, I was transferred to the carrier hospital and here I have been told by the Captain that they captured my boat and prevented it from sinking.*

Signed,
Harald Lange

■ ■ ■

Sweltering in the holding pen below the flight deck, the *U-505* prisoners of war come across some strange markings. They are Morse code dots and dashes scratched into the prison bars by the crew of *U-505*'s sister boat, *U-515*. They tell the story of what befell *U-515*. Apparently, the sub had been sunk by American forces in a battle on April 9, 1944. That's just four weeks ago. Forty-five men had survived, including their commander, Werner Henke.

Grim news indeed. *U-515* is no more! The men of *U-505* vow to resist any pressure to collaborate with the enemy. They will show the Americans what German sailors are made of.

One by one they are brought before Gallery for interrogation. *Guadalcanal*'s ship's doctor, Dr. Henry Morat, serves as interpreter. Earl Trosino also sits in, as does the sergeant-at-arms, just in case any of the prisoners try some funny business. Gallery digs for information that may assist in salvaging *U-505*, but the Germans all offer up the same nonsensical story: Their duties on board *U-505* consisted of emptying out the shit bucket, which Trosino finds especially irritating, as he was the last one to empty it.

One impassioned POW yells: "I'm a soldier of the German Reich. You may betray your country—I will not betray mine! Heil Hitler!"

"Get him the hell out of here," Gallery shouts.

Routine Nazis and liars, one after another, until a tall and very slender prisoner is shown in. He says he is not German. He is Czech.

He learns over Gallery's desk. "I would like to talk to you," he whispers. "If I say something, they will take it out on my family. I cannot talk. I've got to protect my family."

He wants to talk, but he can't talk? *Hmm.* Is this a crack in the German code of silence? Maybe not all these prisoners are unrepentant Nazis after all. It's not much to go on, but Gallery now realizes there may be other non-native Germans among the crew.

And he knows just the guy who can kick that code of silence wide open.

■ ■ ■

Leon "Ski" Bednarczyk was born in New Haven, Connecticut, in 1918. His parents came from Warsaw, landing at Ellis Island in 1901 in search of the American dream. Ski's first language was Polish, which he spoke fluently with his parents at home. He's a big guy, standing just under six feet. In high school he was on the weightlifting team and worked out with his brothers. His hands are enormous, his pinky ring a size thirteen. In 1939, following the German invasion of Poland, Bednarczyk dropped out of technical school and enlisted in the Navy. His intimidating presence made him a natural fit for the military police, and he was assigned to the USS *Guadalcanal* as master-at-arms, running law enforcement and security.

Ski and Captain Gallery have a good rapport. Gallery asks Ski to walk with him on the deck. They can see *U-505* right behind them, under tow by the *Guadalcanal.*

"Ski, we're losing the sub."

Gallery explains that he won't allow Commander Trosino to fiddle with the submarine's diesel engines because he's worried somebody might turn the wrong switch and sink the boat. Without the diesels running, they can't charge the batteries. *U-505* is running out of juice. There isn't enough power to operate the pumps and suck the water out of the boat.

Gallery has an idea that he wants to run by Ski. It seems that among the crew of *U-505*, there are a handful of non-Germans who have been involuntarily conscripted into the Kriegsmarines. Could Ski determine if there are any Poles in the holding pen? Perhaps, Gallery suggests, Ski can turn one of them.

"Let me see what I can do," he tells Gallery.

He heads to the holding pen, stares at the haggard German captives, and eyeballs one POW in particular with classic Slavic physical features. *Betcha he's a Pole.*

"Does anybody here speak Polish?" he asks.

A moment of hesitation. But just a moment. "I speak Polish."

Yep. It's that prisoner Ski had picked out of the rabble that he's pigeonholed as a Pole.

Ski calls him over, unlocks the cage, and tells the guy to come with him. All the other POWs wonder what's going on.

The prisoner follows Ski to the master-at-arms office, where they sit across from each other at a table. His name is Ewald Felix. He is twenty-one and explains that he is half-Polish on his mother's side. He grew up on a farm in Upper Silesia, technically German territory near the border with Poland, but he considers himself more Pole than German. He blames the Nazis for the deaths of his mother and his uncle. He is a junior seaman, having served on *U-505* on just two war cruises.

Ski has a crazy idea. He pulls the sidearm pistol out of his holster and slides it across the table.

Speaking in perfect Polish, he tells Felix: "Go ahead. Want to take a shot at me?"

Felix is stunned. He leans back in his chair, away from the gun. "Why would I want to do that? I'm not part of this."

Ski returns the weapon to his holster. This fellow passed the first test. They chat more. Turns out, Felix hates the Nazis, or so he says. He was compelled to serve on a U-boat under penalty of imprisonment. He says he is nothing but cannon fodder to these Nazis, another expendable Pole. The Germans never trusted him. In fact, when *U-505* shipped out of Lorient, he was never permitted to go topside for fear he might somehow signal the French Resistance.

"Do you know engineering?" Ski asks.

Felix says his rank is apprentice machinist. Yes, he knows how to work the machinery on *U-505*.

"Would you help us blow out the ballast?"

Felix agrees.

Ski tells Felix that he will be returned to the brig, but in one hour's time he is to fake an illness—stomach pains. Leave the rest to Ski Bednarczyk.

An hour passes. On cue, Felix lets the guards know that he is sick and needs a doctor. Ski is summoned to the holding pen and finds Felix doubled over in agony. *Nice acting job*, he thinks. He pulls Felix out of the cage as the other POWs watch.

The next thing, Felix is issued a pair of dungarees, regulation Navy blue shirt, and a white Dixie cup Navy hat. If any Germans happen to see him, at least they won't take him for a POW.

Now Felix is brought before Earl Trosino, who finds the entire scheme shady. With Ski Bednarczyk translating, Trosino asks about technical

submarine matters. Felix seems to be knowledgeable. But why is he bow-
ing his head? He won't look Trosino in the eye.

"Relax," Trosino tells him. He offers Felix a cigarette. Felix is unsure
what to do and Trosino tells him, "Go ahead and light it."

Felix expresses wonderment over this simple courtesy. "They don't do
that in the German Navy."

As Trosino continues the questioning, Felix still won't make eye
contact.

"Tell him when I ask a question, I want him to look in my eyes and I
want to look in his."

Ski translates.

"*Nein. Nein.*" Felix shakes his head and explains, "If I looked in an
officer's face, they would slap me."

"Over here, you're on an American ship. We're a democracy. We don't
do that. You are equal to me. I'm equal to you. Just because I've got
stripes here, that doesn't mean that I'm different."

Felix is also introduced to Captain Gallery. If Felix will help salvage
U-505, Gallery promises the young man he will spend the remainder of
his captivity at a luxury hotel in the United States—unsupervised.

It's a huge gamble, but Gallery agrees that Trosino and Ski should
take Felix over on a whaleboat and see what assistance the *U-505*
mechanic can offer.

Off they go. Gallery pulls Ski aside. He's worried that Felix will do
something to sabotage the U-boat and return to the brig a hero to his
fellow POWs.

"He's not to leave your side," he tells Ski.

Now on board *U-505*, Felix proves very quickly that he knows his way
around the submarine. On his own initiative, he flicks on the bulkhead
fans. The sudden noise scares the daylights out of everyone, but they
relax once fresh air starts circulating inside the vessel. The temperature
cools down and stench starts to dissipate. So far, so good.

Next, Felix shows Trosino how to operate the bilge pump. A hose is
attached and snaked out of the conning tower. Trosino throws the switch.
For forty-five seconds or so, nothing happens. All at once, the hose rears

like a cobra. Water starts spouting out, forty feet into the air. They can actually feel *U-505* rising to the surface as the bilge empties.

Watching from the *Guadalcanal* hangar deck, the sailors give a great big roar of jubilation, so loud Trosino will later say he wouldn't be surprised if they heard it "clean back to the United States."

Trosino convinces Gallery the time has come to speed up the tow. It's another huge risk that could end with the towline snapping again, like it did the night of June 4. As the *Guadalcanal* cranks up to ten knots, the forward speed turns the submarine's propellers. A chain reaction comes to pass. The rotating coils of the electric motors start spinning, which charges the submarine's electric batteries. Finally, there's enough current to run the pumps and dry out the boat. *U-505* achieves full surface trim for the first time since her capture.

The day's hard work at an end, Felix is transported back to the *Guadalcanal*. He spends the night in the master-at-arms room, with Ski. Despite the growing trust he has in the Pole, Ski sticks to his orders and doesn't let Felix out of his sight.

The next day, Felix returns to *U-505* with Trosino and Ski. Now the focus is on weaponry. Felix's attention is drawn to a spare torpedo that's stored in a canister on the top deck. The canister had been damaged during the June 4 battle, exposing the torpedo to the sun.

"You know," Felix says, "if we don't keep that covered and cool, it will explode."

"Uh-oh," Trosino mutters.

Without delay he assigns a sailor to stand over the torpedo with a bucket of cold water, basting it like a turkey. Trosino organizes the sailors to lend a hand. They slide the torpedo on to a dolly and roll it to the farthest point on the fore deck. Then Trosino distributes four-by-four pieces of lumber.

"Now, nobody talks," Trosino says. "Just listen, and when I give the word 'heave,' that's when you heave."

Trosino peers over the U-boat. He's got to time it perfectly. He's waiting for the next wave to slap against the hull at its highest peak. Here comes a big one.

"Heave!"

Over the side the torpedo goes, canister included. Just his luck, a rogue wave appears and carries the torpedo back to the sub. Trosino watches with mouth agape, certain the bomb is going to smash into the hull and blow them all up. But the next moment, the torpedo vanishes from sight and sinks into the ocean.

Felix and Ski follow Trosino down the hatch where they take a quick break to feast on a case of German dill pickles that they discover in the mess hall. Then they head for the aft torpedo room, the site of so much drama with Captain Gallery.

"Be careful around those torpedoes," Felix warns. He's pointing to the acoustic torpedoes. "Don't drop anything."

Felix explains that an acoustic torpedo must be "bled" daily. If the valve isn't released, the pressure builds and the torpedo could detonate by means of "sympathetic sound."

Trosino can only shudder as he recalls pounding on the watertight door with a sledgehammer. He realizes now he could very well have triggered an explosion.

That night, Trosino makes his report to Gallery on the *Guadalcanal*. He expresses utmost faith in Ewald Felix.

"He's as loyal as any man in the United States Navy."

Gallery stares at him. This is a heck of an endorsement. Then Gallery hits him with a tough question.

"Would you be willing to spend the night on board?" Meaning *U-505*. Trosino knows where Gallery's coming from. They've gotten this far; nobody wants to lose the submarine now.

"Yeah," Trosino replies without hesitation.

He rouses Ski Bednarczyk and Ewald Felix. Together, this trio is back to *U-505* for the night.

Trosino intends to get some shut-eye in the captain's cabin, until Ski fills him in on a conversation in Polish he's just had with Felix.

"You know," Ski says, "one of the former captains committed suicide."

"Oh?"

Felix pulls the thin mattress aside and reveals a stain of dried blood and decayed human flesh from that day in 1943 when Peter Zschech blew his brains out. He shows Trosino and Ski a splatter of blood and human remains on the wood paneling.

Trosino can only wonder why the cabin hasn't been thoroughly cleaned. He takes a deep gulp. He's seen enough. Nothing could make him sleep in the captain's cabin. Instead, he spends the entire night awake, working on *U-505* to ensure its stability.

By now, Felix and his new American friends are developing a tight fellowship. Felix eats and bunks with the *Guadalcanal* sailors and everyone seems at ease around him. Felix even asks Ski to make a special request to Captain Gallery: Could he, Ewald Felix, most recently apprentice mechanic on *U-505*, switch sides and sign up with the United States Navy? If so, Felix promises to remain for the "rest of his life and never ask for pay."

When Gallery hears the request, he tells Ski, "No way can I do that. He's still a POW. But I will do everything in my power when he comes ashore. Tell him if there's anything in my power I can do for him I will. I can't express enough how much he's done for this country."

Meanwhile, the rest of the POWs sweating it out in the holding pen are wondering: Where the hell is Ewald Felix? The Americans need to cook up a good story to account for Felix's disappearing act from the other POWs.

As the most fluent English speaker on *U-505*, Hans Goebeler is brought to Gallery's wardroom. Gallery comes right to the point: Ewald Felix is dead from tuberculosis. Or maybe a stomach ailment. In any event, Felix has been buried at sea. Goebeler has authorization to inform his fellow Germans. Perhaps, Gallery suggests, they would like to hold a Catholic funeral for their fallen comrade.

Goebeler can see through this preposterous story with his eyes closed. He knows that Felix is as "healthy as an ox." And isn't it odd that no German delegation had been asked to be present for this secret burial ceremony? If Felix really is dead, Goebeler warns Gallery, "Someone will be held accountable after the war."

When Goebeler returns to the sweltering holding pen, he informs his fellow prisoners about his conversation with Captain Gallery. It doesn't take long for the POWs to put two and two together. Felix is *not* dead—they are certain of that. He can't be found because he is collaborating with the Americans. How else can anyone explain why *U-505* is suddenly floating higher in the water? No doubt about it—Ewald Felix is a stool pigeon and a traitor. They vow to make him pay for this treachery.

■ ■ ■

In Berlin, nothing has been heard from *U-505* since May 15. She should have commenced her return to Lorient by now. A radio transmission ordering Harald Lange to report his current position has gone unanswered. Nothing more is known about the submarine's whereabouts. She is presumed lost at sea.

Of course, the Germans have more far-reaching matters to deal with than the loss of a single U-boat.

On June 6, 1944—two days after *U-505* is seized—a proclamation from General Dwight Eisenhower trumpets the launch of D-day.

Soldiers, Sailors and Airmen of the Allied Expeditionary Force! You are about to embark upon the Great Crusade, toward which we have striven these many months. The eyes of the world are upon you. The hopes and prayers of liberty-loving people everywhere march with you. In company with our brave Allies and brothers-in-arms on other Fronts, you will bring about the destruction of the German war machine, the elimination of Nazi tyranny over the oppressed peoples of Europe, and security for ourselves in a free world.

The sailors comprising Task Group 22.3 in the mid-Atlantic may be far from the landings at Normandy Beach, but they're feeling pretty smug about their own accomplishment, dwarfed though it may be by the invasion of Europe.

"Boy, oh boy," cracks one smart-ass sailor. "Look what Eisenhower had to do to top us."

THIRTY

USS *NEMO*

As dawn breaks, Captain Gallery stands on the deck of *Guadalcanal* and gazes at his prize. He needs to reassure himself that the towline is holding and the submarine hasn't sunk overnight. Towing *U-505* reminds him of a dachshund being dragged by an elephant.

"Well, *Junior* is still with us today," he proclaims with relief.

The day after the D-day invasion, *Guadalcanal* and the task group rendezvous with the USS *Abnaki*. To safeguard secrecy, all the commander of the *Abnaki* has been told is that he is to rendezvous with *Guadalcanal* for a "towing job." He assumed that the aircraft carrier has run out of fuel. Imagine his shock when he realizes that he's expected to tow a captured U-boat!

On board *Abnaki* is Lieutenant Horace Mann, an experienced submariner. He has been issued top secret sealed orders, which he is permitted to open only upon the rendezvous with *Guadalcanal*.

"Take charge of the first thing you see," it reads. That's all.

Mann looks out at the sea. There's the USS *Guadalcanal*. He does a double take. He's being given command of an aircraft carrier? Then he does a triple-take when he sees that the *Guadalcanal* is towing a German U-boat. Okay, now he gets it.

A German U-boat! Mann was expecting a routine tow job, and now *this*! Leave it to Navy Intelligence to spare the interesting details.

They get down to business. Gallery passes the carrier's two-and-a-quarter-inch tow wire over to the ocean tug. To do so, he must bring

Guadalcanal to a full stop, and when she does *U-505* loses her headway and starts sinking again. It's the same alarming setback Gallery has had to contend with since seizing the submarine. Once more, the submarine's bow pops out of the water, the conning tower is awash, and the stern is underwater.

Everyone hustles to make the switch. By 8:00 p.m., the *Abnaki* completely takes over the towing operation and *U-505* is back "in the bag." The crisis over, the U-boat is now underway to Bermuda, escorted by *Guadalcanal* and the destroyer escorts. Meanwhile, the fleet oiler *Kennebec* appears over the horizon. She's a fat and ungainly tanker, but to Gallery she is the most beautiful ship in the Navy. All these days of anxiety about his fuel supply are over at last. Gallery keeps *Gaudalcanal*'s place in formation, and when it's his turn he comes parallel to *Kennebec*, connects the hoses, and begins the process of topping off his nearly depleted tanks. The carrier gulps down a long swig of fuel that should sustain her on the long journey across the Atlantic.

A third Navy vessel, the carrier USS *Humbolt*, out of Casablanca, also rendezvous with *Guadalcanal*. On board the *Humbolt* is Commander Colby Rucker, a qualified American submarine commander, sent by Admiral Ingersoll to lend his expertise.

Gallery tells Trosino to bring Rucker over to *U-505* and show him around. Who knows—maybe he can help.

By now, Trosino considers *U-505* to be his baby, and he's plenty pissed about Rucker's presence. When the tour is over, and Trosino and Rucker are back on *Guadalcanal*, Gallery sends for his chief engineer.

"How'd you make out with Commander Rucker?"

Trosino lets it rip. "Cap'n, I've got something to say and I don't know how to say it. But that man came over there and everything I pointed out, I said, 'We did this, we did that.' And his answers were, 'That's what I would've done. That's what I would've done.' He didn't tell me one thing that was of any use."

"Absolutely nothing?" Gallery asks.

"Absolutely nothing."

Then Trosino offers Gallery some enlightenment about protecting their mutual interests, not that Gallery needs any lessons in Navy politics.

"If you keep him here, he's going to get the credit and what the hell will we have? *Nothing.* I don't need him. *We* did all this. *We* saved it from going down."

Gallery must have been paying attention because the next thing Trosino hears, Rucker is on his way back to Casablanca.

■ ■ ■

Fleet Admiral Ernest King is ready to erupt.

He is holding in his hands a top-secret dispatch from the First Sea Lord of the Admiralty. The dispatch is labeled, *Ultra.* It is addressed to King and marked "personal."

IN VIEW OF THE IMPORTANCE AT THIS TIME OF PREVENTING THE GERMANS FROM SUSPECTING A COMPROMISE OF THEIR CYPHERS, I AM SURE YOU WILL AGREE THAT ALL CONCERNED SHOULD BE ORDERED TO MAINTAIN COMPLETE SECRECY REGARDING THE CAPTURE OF *U-505*.

The First Sea Lord of the Admiralty is the highest-ranking officer of the British Royal Navy. It is an exalted position, steeped in British history, traceable to the year 1689. Currently the title is held by Andrew Browne Cunningham, first viscount Cunningham of Hyndhope, KT, GCB, OM, DSO. Fittingly, with all those abbreviations after his name, Cunningham is better known as ABC, for his initials.

With this dispatch, Cunningham is reminding King that the seizure of *U-505* threatens to render worthless Operation Ultra, perhaps the war's greatest secret, next to the development of the atom bomb. British codebreakers cracked the Kriegsmarine code in 1942. If the German high command learns that one of its submarines has been seized, along with the code books, they'll ditch the current code books, enhance Enigma machine security, and in doing so blind Allied codebreakers. The harm to the Allied war effort could be catastrophic. If not handled with utmost secrecy, the capture of *U-505* could turn out to be a debacle.

The short-tempered King lashes out at the most obvious target—Captain Dan Gallery. King even threatens to court-martial Gallery and only simmers down after coming to the realization that the capture of *U-505* is a triumph—assuming it can be kept under wraps. He also learns that Gallery has kept top brass appraised of his plot—King's own chief of staff, Rear Admiral Francis "Frog" Low, and Commander Kenneth "the Soothsayer" Knowles, who oversees anti-submarine intelligence, had been aware of Gallery's ballsy scheme since March. Apparently, they neglected to inform King, which may have been the source of his fury.

King has a lot of regard for Gallery. He's a trailblazing naval commander and aviator. A true leader. But he's also a maverick who plays by his own rules.

Insulted by the First Sea Lord of the Admiralty's implication that he needed a lesson in preserving the security of Operation *Ultra*, King bangs out his own top-secret dispatch to Cunningham:

PERSONAL FOR FIRST SEA LORD.

BEFORE RECEIPT . . . I HAD TAKEN INITIAL ACTION TO PRE-SERVE SECURITY AND EXPECT TO TAKE FURTHER ACTION TODAY.

Two minutes later, King sends a top-secret dispatch to Gallery and to Admiral Royal Ingersoll, commander in chief, Atlantic Fleet.

ENSURE ENEMY DOES NOT LEARN OF *U-505* CAPTURE.

EMPHASIZE TO ALL CONCERNED NEED FOR ABSOLUTE SECRECY REGARDING CAPTURE.

On June 7, Admiral Low takes it up a notch, recommending that the name *U-505* never be uttered again. Based on suggestions put forward by Commander Knowles, Low offers two "appropriate cover names," and leaves it up to King to make the selection.

TWO SUGGESTIONS FOR YOUR CONSIDERATION ARE:

A) ARK

B) NEMO

King selects *Nemo,* which is Latin for "no one" or "nobody." Of course, it is also the name of the fictional character Captain Nemo from the Jules Verne novel *Twenty Thousand Leagues Under the Sea.* Turns out, Knowles, a former English teacher and magazine editor, is a huge Jules Verne fan.

U-505 is now the possession of the United States Navy, and her newly commissioned name is the USS *Nemo.*

■ ■ ■

As Task Group 22.3 approaches Bermuda, the time has come to take down the swastika flying below the American flag on the jury-rigged submarine flagpole. All evidence that the submarine being towed by the *Abnaki* is of German origin must be erased, to the extent possible.

Gallery calls for twenty-four-year-old Signalman Don Carter.

"Tomorrow morning, I want you to take down the flags, because I don't want anyone seeing it accidentally and know it's captured. We don't want that."

The next day, Carter boards the USS *Nemo* and lowers both flags—and promptly tucks the swastika inside his shirt. While he's at it, he also climbs down the hatch to sneak a peek. He finds there aren't many spoils of war left, because *U-505* has pretty much been stripped bare by previous boarding parties. But he does swipe a pipe, two pens, and a pencil. That's not all: a deck of cards, a pack of cigarettes, and a tiny German-English dictionary. Looking through the dictionary, Carter sees English phrases such as: "Where is the hotel?" And, "Is there a restaurant nearby?" He assumes it was distributed to the German sailors in the event the U-boat sank, and the crew somehow ended up in enemy territory.

When he returns to the *Guadalcanal*, Carter hands over the American flag but stuffs the rest of the booty into his locker, including the Nazi flag.

He should have known better. Two days later, the officer of the deck comes looking for Carter.

"Cap'n Gallery wants to see you."

Carter gulps and heads for the captain's stateroom. Gallery is waiting for him.

"Come in and sit down," Gallery says.

"Yes, sir."

"We want to know what happened to the German flag."

Carter is one of those eager-beaver Americans who enlisted on December 8, 1941, the day after Pearl Harbor, but it was a complicated process. Four hundred young men volunteered that morning at his local Navy recruiting station. Carter was told to go home and stand by. He checked in once a week. Finally, in February, he was sent to Nashville to be sworn in. Because there were no barracks, the Navy took over a hotel to house the men. Nor was there a mess hall, so the Navy arranged for the recruits to eat three meals a day at Mom & Pop's Cafe downtown. After three weeks, those whose names began with the letter A through K were sent to San Diego, and L through Z to the Navy base in Norfolk, Virginia. Carter took the train to San Diego, where he got his vaccination shots and did plenty of marching. There weren't enough uniforms to go around—all Carter was issued was a hat. That's all they had as the United States geared up for world war. A few days later came a shirt. At last, a regulation pair of pants. It wasn't until then that he started looking like a US sailor. Carter was sent to Signal School where he learned semaphore flag communications and the dit-dot-dash of Morse code.

Now he is on the *Guadalcanal*, a participant in a grand adventure, and about to get reamed out by the skipper.

Carter knows better than to bamboozle Gallery.

"Well," he admits, "I've got it. It's down in my locker."

No eruption from Gallery. "Well, I want it," he says calmly. "Don't think I'm going to take it away from you and keep it myself. I want to present it to the Naval Academy in Annapolis. I want it there."

Absconding with a few *U-505* trinkets like a pen or pencil, you can look the other way. But the swastika that flew over the first enemy vessel to be captured since the War of 1812, that's a big deal. Such a flag is an artifact of history that belongs in a museum.

"No problem," Carter mumbles, thankful he isn't going to land in the brig.

Carter returns with the German flag and presents it to Gallery with a sheepish shrug.

"Well, I'll tell you," Gallery says. "I'm not gonna take this flag and not give you something."

With that, Gallery hands over a flare gun taken from *U-505*. Carter looks at his trophy, which is a large bore weapon that fires a shell in the event a ship is in distress. Carter appreciates the gesture, because Gallery didn't have to give him anything. He marches out of there, feeling relieved he didn't get his head handed to him. Of course, he mentions nothing about the pipe, two pens, pencil, deck of cards, pack of cigarettes, and so on, which all remain safely stored away in his locker.

Never underestimate the resourcefulness of the American sailor when it comes to souvenirs.

■ ■ ■

Gallery wants to personally thank the heroes from the *Pillsbury* boarding party. The afternoon before reaching Bermuda, the sailors are transferred to the *Guadalcanal* by breeches buoy. "Flags" Hohne finds the jaunt from ship to ship a little scary. The line looks flimsy. He tries not to look down from his perch.

Once everyone is safely on the deck of *Guadalcanal*, Captain Gallery shakes their hands and extends his appreciation. They all have their photos taken. Lieutenant David offers Gallery insight into what he was thinking as he climbed down the hatch of *U-505* for the first time, not knowing what fate awaited him. David says it reminded him of the biblical tale of Jonah—now he knows how Jonah must have felt when he was swallowed into the belly of the great mammal.

The boarding party is given the run of the aircraft carrier. After all these months on a destroyer, exploring a large ship like the *Guadalcanal* feels like being back on land. Mack Pickels heads to the carrier's soda fountain. Word gets around *Guadalcanal* that the *Pillsbury* boarding party is on board, and their fellow sailors start saluting them for a job well done. The *Pillsbury* sailors spend the night on the *Guadalcanal* in preparation for their arrival in Bermuda the next morning.

But dozens of swastika patches and other trinkets snatched from *U-505* are circulating around the ships, and it's damn challenging to persuade all those sailors to give them up.

Gallery must make them understand that even a *U-505* pen could find its way back to Berlin. All it takes is for one gin-soaked sailor in some bar in Casablanca to boast to a German spy about the time he seized a U-boat and show off a pen inscribed with *U-505* as proof. The triumphant secret mission would no longer be a secret.

On movie night, before the show lights are dimmed and the show gets underway, Gallery steps before his lads. He reminds everyone that regulations require all captured items and souvenirs, no matter how small or trivial, must be handed over to the Office of Naval Intelligence in Washington.

"There is no use whatsoever having a souvenir unless you can show it around and brag about it. So, all those having souvenirs, turn them in tomorrow to the exec's office and no questions will be asked. If souvenirs are found in anyone's possession after tomorrow, no questions will be asked, either—but the boom will be lowered!" Gallery promises that everything will be returned to them once the mission has been declassified.

The next day, the exec's office is "inundated" with what Gallery calls the "damnedest collection of stuff" he'd ever seen—Lugers, flashlights, cameras, hats, German cigarettes, and more. He can't believe that the fellows who risked their lives boarding and salvaging *U-505* could take the time to go souvenir hunting.

Phil Trusheim, the coxswain who piloted the *Pillsbury* whaleboat, grudgingly turns in his *U-505* pistols and binoculars—"nice stuff." They are tagged with his name, and he's assured he'll get them back after the war. But Trusheim doubts he'll ever see his haul again.

Same with Zeke Lukosius.

"They took 'em all away from us," he laments to Mack Pickels.

Pickels manages to squirrel away the pen that he found in the U-boat captain's cabin.

On the *Chatelain*, radioman Joe Villanella is determined to retain possession of a souvenir that's in a class by itself. It's a pair of German-made underwear, stitched with the name "Reh." It belongs to one of the POWs. Villanella found it lying on the deck after the Germans were issued clean clothes on the first day of their capture. Villanella doesn't care if he's brought up on charges—he's hanging on to that underwear. Maybe one day he'll return it to this German fellow named Reh, should their paths cross again.

On board the USS *Pillsbury*, Flags Hohne wonders if he should surrender his Luger and Mauser. He's mulling it over on the bridge at midnight, when a notion strikes him as he stares up at the main mast and gazes at Old Glory, fluttering in the brisk wind. Perfect. The next moment, he lowers the flag and replaces it with another. The old switcheroo! No one will ever know. Then he stores the purloined flag in his locker. Now he's got possession of the flag that flew over the USS *Pillsbury* when *U-505* was captured. They don't call him "Flags" for nothing.

Ritzdorf, the *Guadalcanal* torpedo pilot, is given the chore of censoring letters to make sure nothing classified is going out regarding *U-505*. It's a challenging assignment. Benjamin Franklin famously said that three people may keep a secret, if two of them are dead. But how about expecting three thousand to keep the secret, when these sailors are bursting with pride and anxious to let everyone back home know about one of the greatest war stories ever told? Even Gallery wonders how he's going to get them to keep their mouths shut.

One letter Ritzdorf is looking over is written by a young sailor to his girlfriend in the States.

"I can't tell you where I am," the letter goes, "but I'm on KP and the onions are just making me cry real bad."

You've got to be kidding. *Onions?* The kid is communicating in code, and you didn't need an Enigma machine to figure out what's he's saying. He's letting his girl know he's on his way to Bermuda, which is known for its sweet-tasting onions. With a firm stroke of a black felt pen, the offending sentence is blotted out.

Enough, already! Gallery fires off a letter to all hands.

The operations which we have conducted since 1100 June 4th have been classified as top secret.

The capture of the U-505 can be one of the major turning points in World War number two provided repeat provided we keep our mouths shut about it.

The enemy must not learn of this capture.

I fully appreciate how nice it would be to be able to tell our friends about it when we get in, but you can depend on it that they will read all about it in the history books that are printed from now on.

If you obey the following orders it will safeguard your own health as well as information which is vital to national defense.

The order concludes in vintage Gallery, all in uppercase:

KEEP YOUR BOWELS OPEN AND YOUR MOUTH SHUT.

Not so funny is the following document, which every sailor on Task Group 22.3 is required to sign:

I, [fill in the name], having had the necessity explained to me for mainte-nance of absolute security regarding the capture of the German submarine U-505, do hereby swear that I will reveal this information to no one until the end of the war unless sooner released to the public by the Navy Department. (No one includes my closest relative, friends, military or naval personnel—even an Admiral unless I am directed by my commanding offi-cer to tell him.)

I fully realize that should Germany learn from any source of the capture of U-505 that the tremendous advantage which we have gained by her capture would be immediately nullified, resulting in a great loss to the United States.

I am also aware that should I break this oath, I will be committing a grave military offense, thereby subjecting myself to a General Court Martial, which court has the authority in time of war to award as punishment the death sentence.

A little harsh, for sure, but message received, loud and clear.

THIRTY-ONE

WHERE ARE MY MEN?

The longest tow in US Navy history—2,525 miles, from the west coast of Africa to the Bahamas—is completed on June 19, 1944, when the *Abnaki* hauls *U-505* (make that the USS *Nemo*) into Port Royal Bay, a secure dock on the southernmost point of the island chain, within eyesight of the US Naval Operating Base.

To obscure her origins, a heavy-duty tarpaulin cover is wrapped around the conning tower, blocking her distinctive German design.

On the deck of the *Nemo* stand the nine heroes from the boarding party, beneath a huge Stars and Stripes fluttering from the flagstaff.

Guadalcanal is right beside *Nemo*—carrying an unusual symbol hoisted on the aircraft carrier's main mast. It's a broom, with the bristles pointed up, symbolizing the ship has reached the end of a successful mission with all objectives accomplished. In other words, a "clean sweep."

The *Guadalcanal* anchors and the fifty-nine German POWs are escorted down the gangway under Marine guard to a boat that transports them to shore. There, they are turned over to the commandant and held in strict isolation at a compound inside the naval base. And what a pathetic lot these prisoners make, each man having lost twenty to thirty pounds due to the unrelenting heat inside the prison cage on the carrier during the sixteen-day crossing. But it is a pleasure stepping on dry land again. Until the Navy can figure out what to do with them, the prisoners will be held incommunicado from the outside world, in clear violation of the Geneva Conventions, but that's the way it is has to be for now. Under no circumstances can Berlin discover that the crew of *U-505* is now in American custody.

Gallery parts ways with Kapitänleutnant Harald Lange. In the end, Gallery found Lange to be a "rather decent sort of chap," not at all like the popular image of a rabid Nazi *junker*. They shake hands and Lange thanks Gallery for the medical care he has received on the *Guadalcanal*. Lange is transported to the Naval Base Dispensary, where doctors will determine if his mangled right leg can be saved.

Straightaway, word starts to leak out. Civilians are already lining the dock, pointing to the strange submarine. That tarp concealing the conning tower isn't fooling anyone.

The buzz begins even before *Nemo* arrives. A lighthouse keeper working at the Gibbs Hill Lighthouse tips off Ford Baxter, a reporter with the *Royal Gazette* newspaper, that a German U-boat is about to be towed into Bermuda. To confirm the story, Baxter calls the US Navy base. Panic ensues. *Who told you?* After Baxter refuses to divulge his source, two Marines are sent to the *Royal Gazette* newspaper offices on Queen Street to bring the Canadian-born reporter in for questioning.

The *Royal Gazette*'s editor erupts. "They can't do that! This is not American territory." He puts in an urgent call to the British governor of Bermuda, Lord Burghley. More calls are made to the powers that be on the island, and when the British Marines arrive, they are told they are "not allowed to lay a finger" on Baxter. In return, the *Royal Gazette* promises not to publish the story.

Another pesky reporter also gets wind of *U-505*, only not from a tip but with his own eyes.

Jim Bishop dropped out of school in the eighth grade. He studied typing and shorthand and in 1929 got a job as a copyboy at the nation's most popular tabloid newspaper, the New York *Daily News*. Then he became a cub reporter for the rival *New York Daily Mirror*. Just two years ago, *Collier's* magazine hired him away and now he is literally staring at the greatest scoop of his career.

It happens as he is interviewing Rear Admiral Ingram Sowell, the crusty commander of Naval Operating Base, Bermuda. Sowell is smoking a fat cigar. His burly mug of a face reminds Bishop of a round and cartoonish homemade bomb, with that cigar the burning fuse. Sowell seems distracted. He keeps squinting out his window at the Great Sound.

"Cigar?" Sowell says distractedly as he opens the desk drawer.

Bishop shakes his head. He's there to do a feature piece about the base for *Collier's*, but looking past Sowell, something out that window is catching his attention.

"Tell me," Bishop says, "what's that rusty submarine doing out there?"

"What submarine?"

"The one out there. To the right of the Jeep carrier."

"Oh, that one. It's just sitting out there until I get some orders for it."

"It doesn't look American."

"It doesn't?"

"No. It doesn't."

Sowell glares at Bishop and rolls his cigar around in his mouth. Then he thrusts his face into Bishop's and glares at him with menacing eyes.

"If you so much as mention that boat when you get back home, I'll see that you never get another story."

"It's German?"

"Not one word," the admiral says evenly. "Not even to your wife."

Bishop knows he's sitting on one of the biggest stories of the war and it's all his. He also knows that he'll never write a word about it until the Navy gives him the go-ahead.

So, that leak is plugged but more spring out at the *Guadalcanal*'s home base in Norfolk. Rear Admiral "Frog" Low scolds Atlantic Fleet.

SUBJECT NEMO IS BEING DISCUSSED IN NORFOLK AREA BY THOSE WHO DO NOT NEED TO KNOW.

Meanwhile, the monumental task of assessing U-boat technology and examining thousands of secret German documents begins.

The Soothsayer himself—Commander Kenneth Knowles—and a team of eighteen investigators, engineers, and specialists from Navy intelligence and the Tenth Fleet gather at the Naval Air Transport Service building at National Airport in Washington. They are each issued written instructions from Frog Low, reminding them that "absolute security concerning NEMO" must be maintained. Each passenger is permitted to carry fifty pounds of baggage on the flight. The twenty-six-seat Douglas C-54 Spymaster transport plane takes off at 10:00 a.m. on July 20 and lands in Bermuda two-and-a-half hours later.

Knowles and his team head straight for *Nemo* to eyeball those acoustic torpedoes and pick apart the submarine's radio shack, radar, and sound equipment for evidence of Nazi technological advancements. The sonar equipment is disassembled and flown to Washington for further analysis.

And the interrogation of the prisoners commences.

While this whirlwind of activity is taking place in Bermuda, Lieutenant Jack Dumford, the young communications officer whom Dan Gallery accused of having "secret agent" aspirations, prepares to brief top Navy brass in Washington. Dumford boards a DC-3 military plane from Kindley Field in Bermuda, along with eleven hundred pounds of top-secret documents and the two Enigma machines found on the German submarine. The goods fill a sea chest and nine bulging mailbags.

Upon landing at National Airport, Dumford's plane is surrounded by armored vehicles. The mailbags and sea chest are transferred to a panel truck as Dumford climbs into the passenger seat and is whisked to naval intelligence in the Old Main Navy & Munitions Buildings on Constitution Avenue.

Dumford steps on to the third floor in the seventh wing—headquarters for F-21, the Navy's anti-submarine intelligence command in Washington.

"Hallelujah!" somebody shouts. "We're all set up for today."

Dumford signs for the nine mailbags and the sea chest, officially conveying possession of the *U-505* treasure trove to F-21. He is also informed that a hotel room has been reserved for him, and that it is time for him to go. Dumford is bewildered. *What am I, a delivery boy?* Apparently so. They definitely want him out of there. He can only surmise that the intelligence officers and Navy WAVES are keyed up to start working on the German naval codes without him looking over their shoulder.

But if Dumford feels slighted by the brush-off, he's dazzled the next day when he's taken to the Pentagon and ushered into the office of the big guy himself, Fleet Admiral Ernest King. Dumford has never seen so many admirals in one room—it's wall-to-wall brass, including Admiral Sidney "Slew" McCain, patriarch of the McCain military family. Dumford finds himself the center of attention as these renowned war leaders eagerly gather around to hear the story of *U-505*'s capture. All he can think about is the extraordinary chain of events that brought him to this place. Here he is, a twenty-five-year-old lieutenant junior grade from the sticks of Kentucky who still speaks with an Appalachian twang—and the highest-ranking officers of the United States Navy are hanging on his every word.

When he is finished, Dumford is told he is "free to go." They ask him if he has a place to stay, and he responds that his dad lives in DC. Dumford's father comes over to the Pentagon to pick him up and then Dumford calls his wife, Lillian, in Kentucky and tells her, guess what— I'm in DC! Lillian hops on the next train to see him, and one can only imagine the restraint it takes for him to keep mum about *U-505*.

While Dumford is enjoying a blissful ten-day shore leave, the Navy's submarine intelligence officers assess the material he's brought with him from Bermuda. It's quickly determined to be the real deal—a gold mine. The US Navy now possesses:

- The current cipher codes for the German U-boat fleet in the Atlantic and Indian Oceans for the last two weeks of June 1944. This means the Allies' codebreakers will be able to read U-boat messages in real time, as fast as the Germans receive them. That frees up thirteen thousand hours of precious decoding computer time in these all-important weeks following the D-day invasion.
- The German Navy cipher that is scheduled to go into effect on July 15, 1944, laying out the Enigma machine settings for all U-boats and Nazi surface ships.
- The German codes that are to go into effect starting August 1, 1944. (The German Navy changes its code about every two weeks.)

- A geographic location chart, reporting the precise positions of the entire U-boat fleet at sea. Prior to *U-505*'s capture, this chart had only been partially deciphered by Allied codebreakers.
- Hundreds of U-boat dispatches with the code version on one side and the German translation on the other side.
- Operating instructions for a newly invented German electronic navigation system.

It's an extraordinary cache, a buffet of Nazi secrets.
Over the next ten months, three hundred U-boats are sunk.

■ ■ ■

In Germany, the families of *U-505*'s crew are grieving.

On July 29, 1944, Hans Goebeler's father, Heinrich, receives a letter from the commander of the U-boat flotilla, informing him "with a very heavy heart" that his son is missing.

In spite of several attempts to establish communications and a certain waiting period, [U-505] did not respond. What happened to the brave crew is unfortunately totally uncertain.

Very Honorable Mister Gobler! [sic]. Even if we do not know how the boat was lost we do not wish to give up the slim hope that the crew is safe. Perhaps they were able to leave the boat on the surface and were rescued by enemy vessels. Research to that possibility is underway with help from the Red Cross . . . I would like to ask you my dear Mr. Gobler to please be very patient. Since it is also possible that you will hear from your son first through POW mail I would like to ask you to notify me at once and send me the post card for verification. Dear Mr. Gobler, you can be sure that I share your sorrows for your son.

Heil Hitler!

A letter is also sent to Frau Karoline Hauser, in the city of Zweibruecken. She is the mother of Josef "Raccoon" Hauser, chief engineer of *U-505*.

Hauser's Iron Cross II Class, a pocket watch with chain, nine Reichsmarks, and other personal effects are retrieved from his locker in Lorient and forwarded, along with this chilly message from the Kriegsmarines:

I would like to point out that by sending you these items we indicate that we have no knowledge about the fate of your son.

Petty Officer Karl Springer's fiancée, Ilse, learns that he is missing when her mother-in-law-to-be calls her to come right over. Frau Springer lives in a village eighteen kilometers away, near the border with Poland. Her letter arrives while she's still wracked with grief over the loss of her husband and her son on the Russian front the year before. Now, a second son is gone . . . or "missing," as the letter says. She hands it to Ilse, who reads those matter-of-fact words saying that *U-505* has not been heard from in six weeks and is presumed missing.

Ilse puts the letter down. She knows her fiancé keeps a photo of her over his bunk on *U-505* so he'd think about her every night when he went to sleep. She also remembers something Karl told her when he was on leave. The submarine service isn't like the Wehrmacht, where there's a degree of ambiguity associated with a missing-in-action notification. No such ambivalence exists with U-boats. That is just the nature of submarine warfare. Either you are alive or you are dead. You are never really missing.

"If ever an MIA notice should come, then you can count on it that I am no longer alive," he once told Ilse.

Sitting at a small desk at her home in Hamburg, Carla Lange, the wife of Kapitänleutnant Lange, opens a letter sent by the mother of Chief Petty Officer Willi Schmidt.

"Do you believe our men are still alive?" Frau Schmidt asks.

Another letter from the young bride of a *U-505* sailor really gets to Carla Lange.

"Please, Frau Lange, tell me honestly—is your husband a daredevil wanting to win the Knight's Cross?"

She's trying to blame her husband! Carla writes back that Harald Lange is not reckless. Far from it. He has been at sea since the age of fifteen. He is an "iron seaman" and certainly not an irresponsible glory-hunter. He would never put his men at risk for the sake of a medal. She assures the young bride, "I have not the slightest doubt that our husbands will come home from the war."

A radio is playing, Carla's daughter and son entertaining themselves in the den. It is the daily broadcast of the German armed forces Supreme Command. The 5th Panzer Division of the Wehrmacht is reporting heavy fighting in Normandy. Frau Lange shuts off the radio. Yet another day has gone by and still no word on the fate of *U-505*.

Usually, eight weeks pass before the families of POWs receive word from the Red Cross that their loved ones are alive. But twelve weeks have now gone by and all is silence. In spite of that, for some curious reason, Carla is certain in her heart that her husband is alive.

■ ■ ■

Shirley Jones has a good life in Bermuda. She is nineteen years old with a bubbly personality, soft brunette hair, and a petite build. She is very fit, having excelled at sports at boarding school in Virginia. Shirley's family founded the Pittsburgh Plate Glass Company, and she lived with her parents at Paget Hall, a waterfront estate modeled on grand Newport manors from the Gilded Age. There are maids and a cook named Queenie, a grass tennis court, squash court, and boathouse. Dinner is formal at Paget Hall. The men dress in a coat and tie, and they rinse their hands in finger bowls before dessert is served. The men retire to the smoking room for brandy and conversation about the state of the war, while the women play hearts, bridge, and mah-jongg.

Her family often invites US Navy officers stationed in Bermuda to Paget Hall for dinner, which is how she met the handsome young man she is currently dating, Lieutenant James Humphries, the supply and accounting officer for the US Naval Operating Base.

Shirley has dual American and Bermudian citizenship and is doing her part in the war. She takes Red Cross courses and is a volunteer nurse. To get to the Navy hospital, she boards the ferry and then hops on her bike for a short ride.

One day she is summoned to Admiral Frank Braisted's office and informed she is being assigned to a special mission that requires all her nursing skills as well as total discretion. Braisted explains that she will be shifted to the Officers' Ward and that one of her patients will be a captured German U-boat skipper. His name is Harald Lange and she is warned not to breathe a word to anyone, not even her parents or her Navy boyfriend.

Of course, she can't keep it a secret from Jim, but she makes him swear not to repeat it to anyone.

Shirley is introduced to Harald Lange. Officially, Lange is not a patient at the hospital. No official medical record of him exists. His right leg has been amputated at the knee, and doctors still have a series of operations to perform on the stump to remove shrapnel.

Shirley's warmth wins him over, but Lange is circumspect, except when he brings up his wife, Carla, about whom he can't stop chatting.

He tells Shirley that he's worried she might think he is dead and remarry. He takes it for granted that she has been notified that *U-505* is missing and that presumably all hands have been lost at sea. If only he could communicate with Carla and let her know that he is alive, Lange laments. He is fuming that the Americans are refusing to abide by the Geneva Conventions.

Lange also confides in the young nurse about another concern: that after the war and his repatriation to Germany, he will be "severely punished" for permitting his submarine to fall into enemy hands.

He's lost his leg, his submarine, and his pride. Shirley Jones can't help feeling sorry for him.

Lange has the run of the dispensary from his wheelchair. He plays chess and checkers with the other patients and mingles with officers and enlisted personnel. He chain-smokes cigarettes and overall is a cooperative patient. His English is excellent—he tells Shirley he became fluent when he lived in New York for a year before the war.

So, he actually had a taste of American life. Shirley finds this puzzling.

"Why did you fight for the Nazis?" Shirley asks.

Lange's eyes turn cold. He tells Shirley he blames Great Britain for the rise of Hitler and for "keeping Germany down." But he disavows ever having been a Nazi. Of course, Shirley Jones could not have known about Nazi Party records establishing that Lange joined the party in 1934.

Lange expresses gratitude for the care he's given. When he hears an American patient griping about the food, Lange gives him a brisk talking-to. He refuses to believe that Germany is losing the war. Propaganda, he insists. Lies. To prove otherwise, the base commander authorizes a shortwave radio so that Lange can follow the war news and hear for himself Nazi Germany's dire plight.

Another concern is tormenting Lange. As far as he can tell, he is the only POW from *U-505* in the hospital. What happened to his crew? Where are the men of *U-505*?

THIRTY-TWO

MEDAL OF HONOR

Bermuda is like a resort compared to the prisoners' cage on the USS *Guadalcanal.* Hans Goebeler likens it to a spa. Werner Reh calls it s*chlaffaria*—a wonderland, paradise. As for the interrogations, no big deal. They're not being tortured. Name, rank, serial number—that's all they give up.

Wolfgang Schiller's interrogation takes place in sick bay where he's being treated for nasty skin irritations and a boil on his arm—an affliction that besets half the Germans due to the excessive heat, moisture, and unhygienic conditions they endured on the USS *Guadalcanal.* For security reasons, his Navy interrogator—apparently a Jewish refugee from Nazi Germany—doesn't give his name, but like Schiller he says he's from the same hometown of Breslau.

"I'm not allowed to say anything," Schiller tells him.

The interrogator shrugs. "I know everything anyway."

For the rest of the conversation, they reminisce about Breslau.

In the perfect Bermuda climate, recuperation comes swiftly. The Germans have access to vitamins, but the best medicine is fresh air and the ocean breeze on their faces. The food isn't bad.

They kill time as best they can while they await communication with the International Red Cross to let their families back home know they are being held captive by the Americans. But it doesn't take long before the realization hits the crew that the US Navy has no intention of notifying anyone in Germany that they are alive.

Naturally, the drama of June 4 is replayed endlessly. How did it happen? Who is to blame for this debacle? A kangaroo court of inquiry is conducted, and Hans Goebeler points the finger at Chief Engineer Josef Hauser. Had Hauser done his duty, then *U-505* would be under the sea right now. A consensus is reached: If anyone is to be held accountable, it is Hauser. Once the war is over and they are repatriated to Germany, surely Hauser will face a court martial.

But Hauser fears for his physical safety now. Americans are no longer the menace. His own men could take him down.

■ ■ ■

Captain Gallery and the *Guadalcanal* depart Bermuda on June 20 for the voyage back to Norfolk with a very special passenger—none other than Ewald Felix, sailing under the protection of the Office of Naval Intelligence, with Master-at-Arms Leon "Ski" Bednarczyk at his side, serving as translator.

What to do with the other POWs becomes a headache that reaches the desk of Admiral King. On July 24, the fleet admiral declares that all *Nemo* prisoners—under "highest security"—are to be shipped to Norfolk in the custody of a Marine detachment.

At 8:30 a.m. on August 1, thirty of the prisoners board the medical ship USS *Rockville*, and the rest are taken to the USS *Brattleboro* and depart for the United States.

As soon as the ships reach open sea, several of the Marine guards start puking from severe seasickness. *U-505* Petty Officer Karl Springer rounds one corner and comes upon two nauseous leathernecks sprawled out on the deck, their rifles slung across their listless bodies. Springer is gripped by a wild idea—the Germans could grab the weapons, take over the ship, and throw all the Marines overboard. It could work. *You took our boat, we'll take yours!* But then what? Where do they go? All chatter about an insurrection is ultimately rejected as foolhardy delusion.

Three days later, the ships land in the United States, but instead of disembarking at Norfolk, where curious eyes might observe them, *Rockville* and *Brattleboro* pull into Newport News, due north of the naval base. Responsibility for the POWs is now handed over to the Army's Prisoner of War Division.

The Germans are deloused, interrogated, and issued black POW clothes. But they're not staying very long.

■ ■ ■

Gene Moore, a hot prospect for the Brooklyn Dodgers baseball team, played catcher on the United States Navy Touring Baseball Team in exhibition games against the Army. Pretty soft duty in a time of war.

But that's all over now. The baseball team has been disbanded and Moore and several of his teammates are now assigned to prison guard duty at the naval base in Norfolk.

One night they are roused from a deep sleep and ordered to get dressed. Then they are marched to a Quonset hut and told to sign documents swearing under penalty of court martial never to reveal what they are about to hear.

"Have a seat, men," says a young lieutenant. "Gentlemen, what I'm about to tell you is top secret. You can share this information with no one. Does everyone understand what I just said?"

They nod in agreement.

"We have captured a large group of German sailors who will be transported to a remote camp in Louisiana. . . . They will be held there, not as prisoners of war, but under a classified program that allows us to deny that we are holding them. They will not be allowed to mix with the other prisoners and they will not be allowed any contact with the outside world, period.

"As far as the Krauts back in Berlin know, they are missing in action and presumed dead. I imagine by now that's what their relatives have been told. I can't tell you more than that now, but you have been selected to guard these men. We will be departing in two hours, so pack your seabag and prepare to move out."

Moore has a question. "Why us, Lieutenant? We're the baseball team. I don't think I have held a gun the entire time I have been in the Navy. I don't think anyone in our team knows how to shoot."

The lieutenant says he will make arrangements for the team to undergo small arms training as soon as they arrive at the POW camp. In the meantime, he tells them, "If any of those kraut-eating bastards try to escape, club them with your baseball bats."

Moore and his teammates report to the airstrip. Next stop: Ruston, Louisiana.

At the same time, the *U-505* prisoners are driven by military truck to the local railroad station. They're going to take the scenic route—by train.

Except it's not scenic for the POWs. For two days the train chugs across the beautiful countryside of the Deep South, but the curtained windows have been nailed shut. It's almost like being back on the submarine. They eat meat and beans out of cans.

One thousand miles later, they arrive at their destination: a POW camp set on 750 acres in Ruston. The camp commandant has been issued his orders:

UPON ARRIVAL AT CAMP RUSTON, LA . . . THIS GROUP SHOULD BE KEPT ENTIRELY SEPARATE AND NOT ALLOWED TO COMMUNI-CATE WITH ANY OTHER PRISONERS OF WAR.

That means no mingling with the two thousand German prisoners already in place at Camp Ruston. Nor is any information about their existence to be reported to the International Red Cross.

■ ■ ■

As the *Guadalcanal* pulls into Norfolk, Leon "Ski" Bednarczyk makes the rounds of the carrier and raises $150 from selling *U-505* souvenirs and trinkets that he never turned in. Then he hands the money over to Ewald Felix. Highly irregular but Ski figures they must do something to show their gratitude. Felix is taken to Fort Hood in Alexandria, Virginia, a secret military installation where high-value German captives are debriefed. The classified site is officially known as P. O. Box 1142, after its mailing address. No one knows what the future holds for the anti-Nazi apprentice mechanic who contributed so much to the American cause. Ski promises to stay in touch.

Meanwhile, the *Guadalcanal* crew disperses for some well-deserved shore leave. Ski takes the train to Washington, DC, with one of his *Guadalcanal* buddies.

"You've got to meet my girl, Wanda, and her sister, Darlene," he tells Ski.

Darlene and Wanda Walker share an apartment in Washington, DC, and work for the Office of War Information, the propaganda arm of the US government. The Walker sisters are among the thousands of young American women who have left their homes and flocked to the nation's capital to assist in the war.

Ski's blind date with Darlene goes well. It's a case of opposites attract. She's Protestant, a farm girl from the American heartland—Chillicothe, Missouri. He's Catholic, from New Haven, Connecticut, the son of Polish immigrants. Darlene has wavy, brunette hair and stands a statuesque five-eight, the tallest girl in her class at Chillicothe High School. She loves to dance. Ski has an eye-catching blond handlebar mustache, a pierced ear, and a large tattoo of the American flag and eagle on his bicep. When he flexes his muscle, he can make the flag wave. Darlene has a big laugh over that goofy trick.

They get to know each other. Mindful of Captain Gallery's edict to "keep his mouth shut" or face general court martial and even the death penalty, Ski doesn't breathe a word of *U-505* or the critical role he played. All he tells Darlene is that he's been part of a big mission, and "we've done something nobody else has ever done."

Shore leave goes by too quickly, but before Ski reports back to duty he tells Darlene, "When I come back, we're getting married."

■ ■ ■

Captain Gallery is also in Washington. He reports to Admiral King and squares things away with his mentor, whose fury over *U-505*'s capture has abated now that there is consensus that Gallery has achieved the Navy's greatest intelligence coup of the war. Gallery gives King a special memento. It's a German-language book discovered in the *U-505* skipper's cabin titled, *Roosevelt's Kampf.* The theme? FDR is bent on world conquest! It's farcical Nazi propaganda, and Gallery thinks King will get a kick out of it.

Then King does something Gallery never expected. At his next meeting at the White House, King presents *Roosevelt's Kampf* to none other than President Roosevelt. The gift definitely appeals to Roosevelt's droll sense of humor. On August 19, 1944, Roosevelt writes a thank-you note to Gallery—whereupon all hell breaks loose.

FROM: COMMANDER IN CHIEF

TO: CAPTAIN D. V. GALLERY

SUBJECT: BOOK TAKEN FROM THE CAPTAIN'S BUNK IN THE GERMAN SUBMARINE *U-505*

1. I have received the book forwarded . . . and I have noted with great interest its remarkable history. It will be added to the Library at Hyde Park and will serve as a lasting testimonial of the enterprise, valor and determination of you and your fine task group.

2. Please extend to all of your command my thanks for your fine service to our country and for your thought of me.

Signed,
Franklin Roosevelt

Admiral "Frog" Low is already preoccupied with keeping news of *U-505* from leaking, and when he gets wind of Roosevelt's thank-you memo to Gallery, he throws a conniption. The president knows about *U-505*? First of all, what happened to never uttering the name *U-505*? It's supposed to be the USS *Nemo*. Who else in the White House knows? And is it true that the president wants to add *Roosevelt's Kampf* to his library at his Hyde Park estate where he entertains foreign and domestic dignitaries? Is this a *joke*? Low, who got his nickname from his days on the swimming team at Annapolis, can't very well chew out Fleet Admiral King, or the commander in chief for that matter. So, he turns his rage on Gallery, who in turn counters that if anyone is to blame it is King.

It's just the excuse Low needs to keep an eye on the maverick captain. In September 1944, three months after capturing *U-505*, Gallery is transferred to the Pentagon and assigned to logistics.

One sailor from Brooklyn comes to bid Gallery goodbye. He, too, is getting detached from *Guadalcanal* for another ship. Because the sailor is a "plank owner"—meaning he has been on the *Guadalcanal* since the day she was commissioned—Gallery invites him into his cabin to wish him well and officially hand him his new orders and his last *Guadalcanal* paycheck.

"Well, son, she turned out to be a pretty good ship after all," Gallery says.

The Brooklyn boy grins. "No shit, Cap'n."

■ ■ ■

After all these years, Gallery will finally have an assignment where he's geographically located close to his wife, Vee, and their three kids on the farm in Virginia. He takes advantage, and during downtime, he drives the tractor and chops firewood. Nevertheless, he's "champing at the bit" to get back to the war.

"There is no greater demotion for a naval officer than to go from the bridge of a combatant ship in wartime to a desk in Washington," Gallery laments. At sea, he is a monarch of all that he surveys. Ashore, he's just another "faceless bureaucrat."

The USS *Pillsbury*, with Albert David, Wayne Pickels, Zeke Lukosius, and the other members of the boarding party, pulls into dry dock for refitting and maintenance, still bearing the scar from her encounter with *U-505*'s diving plane. The accolades begin—but only in secret. At the Brooklyn Navy Yard, the Navy Cross—the Navy's second highest medal—is awarded to Lieutenant David, Arthur Knispel, and Stanley Wdowiak. As the first three men from the boarding party to have climbed down the *U-505* hatch, the Board of Awards in Washington has determined that they put their lives most at risk. Silver Stars are awarded to Pickels, Lukosius, Hohne, Jacobson, Riendeau, and Mocarski. Mocarski is also bestowed a Purple Heart, but refuses to accept it because he still has his arms and legs. Silver Stars also go to the three sailors who operated the *Pillsbury*'s whaleboat—Trusheim, Beaver, and Jenkins.

But the existence of *U-505*—or *Nemo*—remains classified. The citations that go with the medals offer no details of their acts of heroism, "for reasons which cannot be revealed at this time."

The two pilots who first spotted and opened fire on *U-505*—Ensign Cadle and Lieutenant Roberts—are awarded the Distinguished Flying Cross.

Gallery is bestowed the Distinguished Service Medal, and Earl Trosino the lesser Legion of Merit, with a *V*, signifying it was earned under combat conditions.

Every sailor on Task Group 22.3 is decorated with the Presidential Unit Citation.

Gallery is really upset. He believes that Earl Trosino is getting the shaft. He had recommended Trosino for the Navy Cross. In Gallery's mind, it's yet another example of the Navy favoring the Pacific theater of war over the Atlantic.

"He did his job in the wrong ocean!" Gallery thunders.

Gallery aims his ire at the Board of Awards and lobbies hard to elevate Albert David's medal as well. It takes several months for the review process to run its course. There is some internal dissent over the fact that Task Group 22.3 did not suffer any casualties in the engagement.

At last, David is notified that he must return the Navy Cross. It has been withdrawn. Instead, he's told he will receive the nation's highest military honor—the Congressional Medal of Honor. He will be the only sailor to win the Medal of Honor for heroism in the Battle of the Atlantic during World War II.

THIRTY-THREE

THEY'RE ALIVE!

The Pentagon has constructed 680 prison camps across America to lock up 425,000 Axis POWs—Japanese, Germans, Italians, and other nationalities that fought for Hitler. Most of the camps are scattered in the remote South and West, where space is plentiful.

The two thousand German POWs incarcerated at Camp Ruston in Louisiana consist almost entirely of veterans of Field Marshal Rommel's Afrika Korps.

Then there are the *U-505* prisoners, who exist in a prison within a prison. They never encounter any of their countrymen. Classified "hard core" or "super Nazis," they are taken to a far-off compound on the north quarter of the vast camp, isolated from the other prisoners by acreage, barbed wire, and a wide, empty field that lies between them.

The *U-505* compound is about half the size of a football field, carved out of the forest, and put together in a frenzy of construction with cheap plywood barracks and flimsy tents. No one would be surprised if a tornado swept it all away.

Machine gun placements have been situated at all corners. Beyond the barbed wire is a swampland inhabited by venomous snakes and hungry alligators. Escape is impossible, and anyway, escape to where? Germany is five thousand miles away. The *U-505* sailors can't even be certain they know where they are. They have no bearings. For all they know, they could be imprisoned on an island. But discipline remains high, and they are aching to somehow get word back to their loved ones in Germany that they are among the living.

A campaign of de-Nazification is ordered by the camp commandant, Navy Commander William "Dirty Bill" Arbeiter. The Nazi salute is banned. So is uttering the words "Heil Hitler." But the *U-505* Germans are an obstinate bunch. When Werner Reh breaks his toe playing a game of soccer, he manages to pull off the Nazi salute as he passes a fellow prisoner, despite walking with crutches. A guard catches Reh and he's given thirty days' punishment—breaking rocks.

Of course, writing home is prohibited. The capture of *U-505* remains top secret.

Tedium is a major cross to bear. To pass the time, the men organize classes in math, art, and English. Meantime, they dig. The tunnel is situated under the barracks. The POWs are full of bluster about burrowing under the fence and making their way to South America. Spanish becomes a popular course of study. None of the prisoners really take the escape plot seriously, and the tunnel is abandoned before they make much headway.

The POWs try other schemes. They mix some cleaning chemicals they find in the kitchen and concoct a batch of hydrogen gas. Then they fabricate balloons out of cellophane wrappings and scrawl "U-505 Lives!" across them. The balloons are sent aloft. Clever bunch. "Dirty Bill" Arbeiter is impressed by their ingenuity. Of course, the balloon ploy yields zero results.

Every now and then, a box turtle crawls across the grounds, leading to the hatching of another preposterous scheme. Swastikas are painted on the shells of the little creatures in the fanciful notion that they'll make their way out of the camp. No, not to Germany, just to the other side of Camp Ruston to let their countrymen know about their presence. Give them an A for effort.

The prisoners are put to work in the Louisiana forest felling trees. It is there that the fearless Hans Goebeler encounters the terror of snakes dropping from tree branches and falling on his head. And then there are giant wasps that swoop down and bite his exposed skin and are so monstrous they merit a nickname—"*stuka*"—after the Luftwaffe dive-bomber. Goebeler has never worked so hard in his life, and all for a paycheck of eighty cents a day.

Extraordinary efforts are made to keep the International Red Cross from discovering the existence of the *U-505* prisoners. When a regularly scheduled Red Cross inspection takes place at Camp Ruston on November 2, 1944, the *U-505* prisoners are bused ninety miles to a military facility at Camp Livingston. All goes well, until the Red Cross delegation demands a tour of that remote compound at the far end of the camp. Obfuscation ensues. The Red Cross is razzle-dazzled by a puzzling claim that the site is "forbidden" to civilians by order of "superior authorities."

"We were denied the visit for reasons which did [not] seem to us to be very clear," they declare in their official report.

Two weeks later, Óberleutnant Paul Meyer, the senior *U-505* prisoner in the absence of Harald Lange, writes a formal complaint to the Swiss Legation in Washington declaring his belief that "their families in Germany have not been notified of their capture," in violation of Article 6 of the Geneva Conventions.

Meyer's letter is never delivered. He follows up with three other letters to the Swiss Legation, all of which are collected by the military police and diverted to the Bureau of Naval Intelligence in Washington where they are stored for safekeeping. No letter ever reaches the Swiss.

In November 1944, with the wholesale surrender of German military units in Europe, space must be found for the influx of prisoners at Camp Ruston. To make room, the commandant orders sixteen German soldiers relocated to the *U-505* barracks.

Big mistake. One of those Wehrmacht POWs is a German named Kurt Geuthner. There he is, checking out his new surroundings, when he encounters a *U-505* seaman, Erich Wilhelm Kalbitz. They know each other! Geuthner's wife is Kalbitz's cousin. What are the odds?

Naturally, Kalbitz tells Geuthner the whole crazy story of *U-505*. Whereupon Geuthner writes a letter to his family relating the astonishing news that their cousin Erich Kalbitz is alive. The letter is delivered to Germany via routine POW correspondence channels and somehow sails right through military censorship in the United States. It's an extraordinary breach of security.

The next month, Kalbitz writes a letter to his family declaring that he is a prisoner of war in America, in a place called Louisiana. He slips it to his cousin Kurt Geuthner, who smuggles it out of Camp Ruston in one of his own letters. In due course, the Kalbitz letter ends up at the US postal censor's office in Manhattan, where an eagle-eyed Navy intelligence officer spots it before it is forwarded to Germany.

An investigation is launched and the Kalbitz-Geuthner connection is ferreted out. Geuthner is transferred out of Camp Ruston, the guards are reprimanded, and security overall enhanced.

■ ■ ■

In February 1945, the time has come for the skipper of *U-505*, Harald Lange, to be discharged from the Navy hospital in Bermuda.

Shirley Jones, the pretty American nurse assigned to his care, accompanies Lange in the ambulance for the drive to the embarkation point where a Navy flying boat is ready to airlift Lange to the States. In a way, she's going to miss him. True, he was the enemy, but the poor man lost his leg, and he may never see his family again.

Lange seems tense. Shirley tries to make small talk. She asks an innocent question. "Which POW camp are you going to?"

Lange turns on her. "How would I know? What a stupid question!"

So very unlike Lange. Shirley wonders if he's still worried whether his wife has forsaken him and remarried. Good soul that she is, the nurse doesn't hold a grudge. As she bids farewell to Lange, she tells him she wishes him well and hopes he'll be reunited with his family again.

■ ■ ■

On April 30, 1945, facing total military collapse on all fronts, Adolf Hitler puts his Walther PPK pistol to his right temple and blows his brains out. His war has cost the lives of an estimated seventy-five million military personnel and civilians—3 percent of the world's population.

Eva Braun, Hitler's wife of forty hours, bites into a cyanide capsule and is found next to the dictator on a small sofa. Their bodies are rolled into a carpet and carried out of the Führer's bunker and burned in the Reich Chancellery's garden as Soviet troops storm into Berlin.

In accordance with Hitler's last will and testament, dated just the day before his suicide, Grand Admiral Karl Doenitz, mastermind of the U-boat campaign, is named chancellor and head of state.

On May 4, Admiral Doenitz sends a signal to the entire U-boat fleet ordering the cessation of attacks on Allied shipping.

"My U-boat men, six years of U-boat warfare lie behind us. You have fought like lions. Comrades, maintain in the future your U-boat spirit, with which you have fought at sea bravely and unflinchingly for the welfare of our Fatherland. Long live Germany!"

The Allied navies are taking no chances. U-boat captains are among the most fanatical Nazis and might fight to the death with their last torpedoes.

In fact, the next day, the American cargo ship *Black Point*, carrying coal to Boston, is torpedoed off the coast of Rhode Island, with a loss of twelve lives. US destroyers and the Coast Guard converge on the location and send the U-boat, identified as *U-853*, into oblivion with a spray of depth charges. All fifty-five German sailors are killed. It is never established if the captain of *U-853* ever received the order to cease fire. *Black Point* is the last casualty of the U-boat menace in World War II.

Doenitz's regime lasts seven days. Germany signs the instrument of unconditional surrender at a little red schoolhouse in Reims, France, that serves as headquarters for General of the Army Dwight David Eisenhower. And so ends the Third Reich that Hitler vowed would last a thousand years.

Victory in Europe is declared.

The world rejoices.

In Paris, sirens wail and cannons boom and the famed boulevards of the City of Light are jammed with jubilant civilians. Fireworks blaze across the night sky. The evening newspapers publish extras announcing the formal capitulation of the German armed forces and the renunciation of Nazi principals. General Charles de Gaulle leads Parisians in singing the national anthem, "La Marseillaise."

In London, Prime Minister Winston Churchill appears on the balcony of Buckingham Palace with King George and nineteen-year-old Princess Elizabeth, the future queen. Churchill gazes at the exultant throng and makes a deep bow, and it seems that all of England responds with a roar. Searchlights sweep the sky forming colossal Vs for "victory." Big Ben shines over the River Thames, illuminated once more after nearly six years of darkness.

In Washington, the Capitol Dome gleams with floodlights for the first time since a blackout was imposed in 1941 following Pearl Harbor. There is also sorrow that Franklin D. Roosevelt could not have lived to see this glorious day. President Harry S. Truman, in office just twenty-seven days, officially proclaims VE Day (Victory in Europe). As millions of Americans across the nation gather around their radios, Truman declares:

"The flags of freedom fly over all Europe. The West is free."

In New York, a million people—civilians, sailors, soldiers, bobby soxers—spontaneously converge on Times Square, kissing strangers, tooting horns and bugles, clanging cowbells, banging drums, and dancing and cheering with joy. The cry goes out, "It's over! It's over!" On Wall Street, the windows are flung open and a blizzard of paper of every conceivable kind—torn out of telephone books, ticker tape, confetti, decks of playing cards, streamers, anything—flutters out of skyscrapers to the concrete canyons below. The celebration is tempered by awareness that the war in the Pacific is yet to be won.

Praise is heaped on General Eisenhower for commanding the fighting forces of the United Nations into a unified war machine that has crushed Hitler's Fortress Europa. In a radio broadcast, Eisenhower proclaims the real hero of the war to be "G.I. Joe."

■ ■ ■

Bosun First Class Wayne "Mack" Pickels is on board the USS *Pillsbury* when VE Day is declared. The destroyer escort is given the mission of accepting the surrender of the first U-boat to turn herself in following the cessation of hostilities. She is *U-858*, skippered by none other than Thilo Bode, who was second in command of *U-505* when Peter Zschech served as captain before his suicide.

After breaking surface about 750 miles off the Eastern Seaboard, Bode hoists a black flag, signaling submission, and radios the US Navy for further instructions. The *Pillsbury* is sent to rendezvous with the U-boat. Navy planes and two blimps fly overhead as Bode and his men are taken into custody and sent over to the *Pillsbury* while a skeleton crew of German sailors remains on board to operate the U-boat with American guards keeping an eye on them. The U-boat's deck guns are dismantled, and her fourteen torpedoes neutralized. She is stripped of all destructive powers. In rough seas, *U-858* sails into the Navy port near Cape May, New Jersey, and Bode is sent to a POW camp in Fort Miles, Delaware, unaware that his old submarine, *U-505*, had been seized almost a year ago.

At all POW camps in America, an Army proclamation is posted:

THE ORGANIZED RESISTANCE OF THE GERMAN ARMED FORCES HAS CEASED TO EXIST. THE NATIONAL SOCIALIST GOVERNMENT OF GERMANY HAS CEASED TO EXIST.

At Camp Ruston in Louisiana, the *U-505* sailors find it hard to believe. Loyal to the bitter end, they had just celebrated May Day with the hoisting of a homemade swastika.

After all this time, they are finally permitted to write home to their families in Germany, but in the chaos of postwar Germany, who can be certain if the letters will ever be delivered?

In Germany, Petty Officer Karl Springer's mother learns the astonishing news via a back channel—Carla Lange, the wife of *U-505* skipper Harald Lange.

"Your son and the crew are doing well," Carla Lange writes in a letter to Frau Springer. "Except my husband lost a leg during the affair."

No other information is forthcoming. It is wondrous news—if it is true. Frau Springer writes back and asks Carla Lange how she can make such an astonishing assertion since the Springer family has heard nothing official.

Frau Lange doesn't like being doubted. Her reply is simple and direct:

"I am the wife of the commander. I have confirmation. I know that the submarine wasn't sunk."

THIRTY-FOUR

NOW IT CAN BE TOLD

Eight days after Germany's unconditional surrender, the US Navy Department announces to the world that "one of the best kept secrets of the war" can now be told.

"CAPTURE OF NAZI SUBMARINE IN 1944 REVEALED," declares the press release issued by the Navy Office of Public Relations.

Now the three thousand sailors from Task Group 22.3 can finally talk about what happened without fear of court martial or the firing squad. As far as anyone knows, there had not been a single leak from any of the American sailors. For Captain Dan Gallery, this is perhaps Codename *Nemo*'s most impressive achievement.

"The boys did keep their mouths shut," says Gallery. "I think this speaks very highly indeed for the devotion to duty and sense of responsibility of the average American wearing bellbottom trousers."

With the capture of *U-505* now declassified, Gallery is given permission to write an article for the *Saturday Evening Post* detailing the "greatest [intelligence] windfall of the war."

Gallery's article is published under the headline, "We Captured a German Sub." He praises Earl Trosino ("uncanny instinct for finding the right valves and total disregard of his own safety") and the heroism of the *Pillsbury* boarding party, particularly that of Lieutenant Albert David and the sailors who followed him down the hatch.

"There was every reason to believe that there were still Nazis below, opening sea cocks and getting ready to blow up the vessel. The very fact that the sub was running a good speed, surfaced, indicated that she was not totally abandoned. But this didn't give David any pause. Without hesitating, he and A. W. Knispel . . . and S. E. Wdowiak plunged down the conning tower hatch ready to fight it out with any krauts below."

Gallery harkens back to the Continental Navy of 129 years ago, when the US sloop *Peacock* boarded the British ship *Nautilus* in the Strait of Sunda, East Indies. Had those sailors from long ago witnessed the events of June 4, 1944, he writes, they would have been bedazzled by the weaponry of modern warfare, now carried out by ships made of steel that maneuver without wind, and mechanical birds that fly in the air, and strange looking vessels that sail under the surface of the sea. But the ancient mariner's rallying cry, "Away, boarders!" is a call to arms they certainly would have recognized.

Gallery's article is a sensation, and overnight he becomes one of the best-known naval officers in America.

Meanwhile, the USS *Nemo* is back in operation—this time under the command of Lieutenant Horace Mann of the US Navy and his hand-picked crew of thirty-two American submariners. It's a tight fit for many of the sailors who are considerably brawnier than their German counterparts. One electrician's mate finds himself assigned to a top bunk with pipes running a few inches above his face. Good luck should he turn over during the night.

Nemo sails to Norfolk under its own power but on the surface only, having been deemed too fragile to submerge. There is one harrowing moment near Cape Hatteras when a fighter plane spots the U-boat and comes in low and fast. What a way to go—bombed out of existence after the war is over because you're mistaken for a renegade German submarine. But the crew scrambles topside and wave the Stars and Stripes. The plane flies low, dips its wings, and heads off into the wild blue yonder.

Nemo's new mission is raising money for war bonds. Step right up! Buy a $25 war bond, and you get to board the world's most famous submarine. Thousands of citizens line up to inspect the vessel once known as *U-505*. Everyone gets a kick out of the recordings played over the boat's public address system—music the German sailors listened to when *U-505* was stirring havoc during the war, wistful pinings for the old country:

"Bells from the Homeland," and "Homeland I Would Like to See You Once Again."

From Norfolk, the submarine proceeds to Savannah; Charleston; Portland; New London; Connecticut; and finally New York City where she docks at Pier 88 on the Hudson River. Five thousand youngsters who pitched in for a paper drive are permitted to climb on board for free. In all, $17 million in war bonds is raised.

Lieutenant David is sent on a publicity tour. He arrives in Manhattan for a round of newspaper interviews at the federal office building at 90 Church Street, and in his self-effacing way, relates the harrowing account of how he became the first American to climb down the hatch.

"We began pulling wires from the time bombs, and every time I pulled one I said, 'Here goes nothing,'" he tells reporters.

David is flown to Madison, Wisconsin, for a reunion with his wife, Lynda, the practical nurse he married in 1942 after a whirlwind romance. He shows Lynda his war booty. Somehow, he managed to smuggle a staggering array of *U-505* souvenirs: a pair of binoculars, a clock, pistol, typewriter, and goggles.

David is informed that he is to receive the Medal of Honor from President Truman on October 5, in the White House Rose Garden. It will be a special occasion: The ceremony falls on "Nimitz Day," when the commander of the Pacific Fleet, Admiral Chester Nimitz, is to be honored. It was Nimitz who accepted the unconditional surrender of Japan on board his flagship, the USS *Missouri,* on September 2, 1945. A ticker tape parade for Nimitz up the Canyon of Heroes in New York City is to follow the next day. At last, the world is at peace in all theaters of war.

Albert David and his wife proceed to Norfolk where they find a little apartment near the Navy base. Hailed a national hero, these are heady days for the Navy veteran. It's time, he decides, to put in his retirement papers after twenty-six years of service. He's also not feeling that great. He has himself checked out by Navy doctors.

Death comes to Albert David not from a *U-505* booby trap but from the ticking time bomb in his own chest. Nineteen days before the White House ceremony is to take place, David suffers a massive heart attack at home. It's a widow-maker. The cause of death is determined to be coronary occlusion due to clogged arteries. He is forty-three years old.

A shaken Lynda David escorts her husband's body by train to Madison, where he is buried. Instead of President Truman at the White House, the Medal of Honor is posthumously presented to the widow at a ceremony held in the auditorium of the Naval Station Great Lakes training center in Illinois. She is dressed in black. The citation honoring her late husband reads, "For conspicuous gallantry and intrepidity at the risk of his life above and beyond the call of duty."

In 1946, the *Pillsbury* sails into Green Cove Springs, Florida, and is decommissioned—in other words, mothballed. The sailors are discharged from the Navy and go their separate ways, resuming their civilian lives from long ago, before the war. Back to the way of life of ordinary Americans making a living and raising a family.

Mack Pickels is let out on Christmas Eve and returns home to San Antonio, Texas, where he enrolls in refrigeration school under the G. I. Bill (formally known as the Servicemen's Readjustment Act). Westinghouse Electric hires him as a refrigerator repairman at a starting wage of $1.16 an hour. One night, he's having dinner at the Barn Door, a popular steak house in San Antonio, where he meets a waitress, Jacqueline Wurth. She's wearing a fetching floral blouse, and her date has stood her up. Perfect. She and Mack get to talking. Turns out, her parents come from Switzerland, and her two oldest siblings were also born in Switzerland.

"Oh," says Mack, "so, you're Swiss."

"Nope. I'm Texan."

The rest, as they say, is history. When the time is right, Mack proposes.

"Let's get married on June fourth," he says.

The marriage is off to an excellent start. Now Mack Pickels knows he'll never forget his anniversary. The capture of *U-505* and his wife's heart, on the same day of the year.

Gordon "Flags" Hohne returns to Worcester, Massachusetts, and marries his childhood sweetheart, Norma. Hohne serves in the Navy Reserves, returns to active duty during the Korean War, and considers making the Navy his career, but Norma puts her petite but unyielding foot down and says, "It's either me or the Navy." Norma wins. Hohne finds work manufacturing heavy-duty chains for the Rex Chain Belt Company.

Zeke Lukosius is discharged in July 1946 and moves back to his old neighborhood on the South Side of Chicago before settling down in the suburb of South Holland. He becomes an industrial roofer, raises four children with his wife, Dorothy, and is an active member of the Veterans of Foreign Wars, Post 9964.

Arthur Knispel heads back to Newark and joins his brother, William, on the Newark Fire Department. Sometimes, Knispel harbors a little resentment because he had been designated to be the first man down the submarine hatch, but Lieutenant David bumped him aside for the noblest of reasons—he thought it was his duty as officer in charge to take the bullet in the event of a shoot-out with the Germans. But Knispel figures he deserves the Medal of Honor just as much as David. After all, Knispel was right behind the lieutenant and took as much risk.

Stanley Wdowiak manages to take leave of the *Pillsbury* with a spoil of war—his pair of Zeiss binoculars from *U-505*. He returns to Brooklyn and later moves to the borough of Queens with his wife, Sissy, and finds a job delivering milk to restaurants in Manhattan. The work requires that he wake at two o'clock in the morning to make his run.

Gunner's Mate First Class Chester Mocarski also makes it back to Brooklyn and suffers back pains for the rest of his life due to the injuries sustained when he leaped off the whaleboat attempting to board *U-505* and was crushed between the boats.

Electrician's Mate Third Class William Riendeau musters out of the Navy on August 2, 1946, moves back to Rhode Island, and becomes an electrician.

Chief Motor Machinist's Mate George Jacobson, at age forty-four the oldest member of the boarding party, opens a motor repair shop in Portland, Oregon, and later manages an apartment house where he lives with his wife and stepdaughter.

The crew on the USS *Guadalcanal* also shoves off for parts far and wide. Before Leon "Ski" Bednarczyk disembarks, Captain Gallery has a chat with him. Gallery explains that Ewald Felix must be protected against Nazi retaliation, and his collaboration with the US Navy must remain confidential.

"Ski, you know I'm not going to be able to write about what you did. It's classified. I can't tell the story. But I want you to know what you did was incredible." Gallery recommends Ski for the Silver Star but the Board of Awards won't go higher than a Navy and Marine Corps Medal. Ski accepts the fact that it's one of those things. He'll never get the recognition he deserves.

Ski did end up marrying Darlene Walker. Darlene is seventeen when they find a justice of the peace in Newport News, Virginia.

"Just put down you're eighteen," Bednarczyk whispers.

More surprises are in store when they make the drive to New Haven in a Model A Ford to meet his parents. Ski forewarns Darlene to fib about her religion and tell his parents she's Catholic.

In the Bednarczyk household, they speak only Polish. All the food laid out on the table consists of traditional Polish delicacies.

"You're going to have to learn to cook Polish food," her groom tells Darlene, if she is to be accepted into the Bednarczyk clan.

Darlene wonders, *What have I gotten myself into?*

Mrs. Bednarczyk gives Darlene a cookbook, *The Olde Warsaw Cookbook*. On page 185, she finds a great recipe for cheesecake.

With World War II at an end, Bednarczyk is trying to find his way, like so many servicemen suddenly flung back into civilian life. He drives a truck. He and Darlene move to her home state of Missouri for a time, and he flies a crop duster. Since every household in America seems to be buying that wondrous invention, the television set, Ski decides to attend trade school and learn TV repairs. Then his old commander throws him a lifeline, and Bednarczyk finds his true calling. With a letter of recommendation from Dan Gallery, he's hired by the New Haven fire department.

Commander Earl Trosino is detached from *Guadalcanal* and sent to the Navy's Damage Control and Fire Fighting School in Philadelphia, close to home in Springfield, Pennsylvania. After all these years at sea, he gets reacquainted with his wife, Lucy. The cheeky commander likes to say of this period of his life, "I was putting out fires at the school, and she was putting out fires in me." Trosino is discharged from active duty in 1946 and returns to sea as chief engineer for Sun Oil tankers.

The USS *Guadalcanal* is decommissioned in July 1946 in Norfolk and put in mothballs. Captain Gallery boards her for one final, nostalgic farewell. Walking the hangar deck, he sees only ghosts. He prefers to remember her as she was in her heyday—a good ship with a valiant crew.

Guadalcanal will end her days sold, wouldn't you know it, to the Japanese. For scrap.

THIRTY-FIVE

"DEAR MOTHER!"

t's time to go home.

Repatriating the POWs is a vast undertaking involving international coordination and unprecedented logistics among the four great powers now governing occupied Germany.

For *U-505* radio operator Karl Springer, the worst part of the war is right now, when he doesn't know how to reach his fiancée, Ilse. At last permitted to communicate with his family, his letters go unanswered. What has become of Ilse? Has she survived the war?

In July 1945, Camp Ruston starts emptying out. Springer is one of the first POWs to depart. He's bused to Camp McCain in Mississippi. A thought strikes him. He wonders if Ilse might be with her brother, who owns an inn somewhere in the Ruhr Valley, but for the life of him, Springer can't remember the town. He can't even remember Ilse's brother's name! Too many years of war have taken their toll.

He's in his barracks at Camp McCain. Springer has taken up knitting, of all things. He sews sweaters out of the threads from cleaning rags. There are fifty other prisoners with him, all wearing khaki shirts and pants with the letters *PW* stenciled on each leg. The food is good, prepared by German cooks. A typical supper is meat loaf and scrambled eggs. Beer is available in the canteen.

Ilse. Ilse.

All at once, like a slap across the face, Springer remembers. *Foreman.* Ilse's brother's name is Foreman! Then, the name of the town pops into his head: *Schwerte*. That's where Foreman lives! He writes it on the wall, so he won't forget. Maybe that's where Ilse has found a haven from the fighting. At least it's a start. He sends Ilse a postcard in care of her brother, in Schwerte. No specific address.

Weeks pass. Remarkably, the postcard is delivered. Isle is holding it in her hands. She can't believe it. Her fiancé is back from the dead.

Springer remains in Camp McCain for eight more months before he's sent to the next stop on the long journey home, Camp Shanks in New York, just twenty-five miles from Manhattan. There he learns that he is to be deployed to England, or possibly France or Belgium, as a farm laborer or to clear land mines.

That does it. Springer joins his fellow POWs on a hunger strike. Of course, such an act of disobedience would never have been tolerated in a Nazi *stalag*. But this is America. In Springer's retelling, after three days, a senior official in the camp informs him, "I give you my word of honor as an American officer, you will not go to England, France, or Belgium. You will go to Germany!" The hunger strike is called off.

Finally, in April 1946, Springer boards a troop transport ship. Where does he end up? In Liverpool! It takes another seven months before he's finally sent to the city of Munster in Lower Saxony and released. Springer boards a train to Schwerte, and when he gets off, there is his fiancée, Ilse, at the station, running into his arms. Springer doesn't even have a "pot to piss in." His only possession is a ten-pound bag of coffee beans that he purchased in Liverpool. But as a qualified electrician, he has a skill in high demand. In just two days he finds work. His new boss even offers a perk—a furnished apartment. On Christmas Eve, he and Ilse are married.

It takes Josef "Raccoon" Hauser, *U-505*'s despised chief engineer, nine months to get back to his family.

Hauser has had a rough time at Camp Ruston. He's still blamed for allowing *U-505* to fall into American hands because he failed to set the detonation charges. Hauser fears assassination at the hands of his own men. For his own protection, he is placed in isolation at Camp Ruston's hospital. Then, to his enormous relief, in February 1946, he's sent back to Europe by way of a US troop transport ship and ends up at Allied

POW Encampment #23, in Bolbec, France, about a ninety-minute drive from the D-day landing at Omaha Beach. Hauser is permitted to write his mother, Karoline, who lives in the American Zone in Bavaria. His words express nothing about the disgrace and anguish he's suffered through.

Dear Mother!

After two years, I am again on the coast of Western Europe. I am healthy, and am returning soon. I have had no news from you, my fiancée and siblings since my last journey. Hopefully you are still alive and well . . . yesterday I got the order for my release. American Zone: Residency Munich. This is my destination after my release, which is supposed to happen very quickly. So we can meet again soon. About everything else we'll talk. Dear Mother! Give my greetings to my lovely fiancée, my siblings, and all our relatives in Munich and wherever they may be scattered.

Hauser's chief accuser, Hans Goebeler, is shipped to England in December 1945. From there he's sent by rail to Edinburgh, Scotland, and a labor camp known as POW Camp #123, located adjacent to an RAF airfield.

He works on a farm for a shilling a day. The hours are long, from 6:30 a.m. to 5:30 p.m., six days a week. He has Sundays off and freedom of movement but resents having been "sold" to the British as a forced laborer even if he has to admit the living conditions are comfortable and he probably has it better off than his family in Germany.

Two years pass. Two years! At last, in December 1947, he is permitted to return home with the understanding that once back on German soil he is to participate in a "de-Nazification" program, designed to purge German society of Hitlerism and Nazi ideology and introduce returning military personnel to the principles of democracy. They are also required to watch films of concentration camps so that all Germans can understand the atrocities committed by the Nazis.

As Goebeler boards a ferry and crosses the English Channel, he is stricken with inner conflict. His family has survived the war, and he is thrilled to be reuniting with them in the coming days, but the idea of de-Nazification deeply disturbs him. He swore an allegiance to the Third Reich and is "revolted" by the thought of swearing an oath to an "enemy-imposed regime."

After landing at Calais, a brooding Goebeler boards a train packed with hundreds of other newly released German POWs. The train chugs its way across Belgium, to the border city of Duisburg, Germany. Goebeler makes his decision. He leaps off just before crossing the frontier. No de-Nazification for this *U-505* crewman. He manages to smuggle his way across the border and from there to the village of Bottendorf, two hundred kilometers away. Home at last.

■ ■ ■

In June 1946, Leon "Ski" Bednarczyk receives a letter addressed to his parents' house on Nash Street in New Haven.

It's from Ewald Felix. He's back in Germany.

"Dear Friend," the letter begins in broken English. "Now it is nearly two years ago that we parted at Norfolk, but I didn't forget going to *Guadalcanal* yet. I hope that you were still remembering me, the guy of *U-505*."

Felix reveals that he has been repatriated to Germany, which has been split into four zones of occupation—American, British, French, and Soviet.

Right now, Felix is living in the small town of Minden, situated in the British military zone. He arrived four weeks ago. He informs Ski that he had a "swell time" in the fourteen months he spent at Fort Hood in Alexandria, Virginia, the secret military installation also known as P. O. Box 1142 where high-value German captives are debriefed. It's also where he picked up conversational English.

But now it's a different story. Like millions of his countrymen in the chaos of postwar Germany, these are hellish days for Felix. There is no work, he tells Ski. He is hungry.

Well, what are you doing now? Have you got your discharge from the Navy or my friend Trosino?

I'd very much appreciate your answering my letter in shortest time possible.

Yours sincerely,
Ewald Felix

It's a shock to all his American patrons that Felix has been sent back to Germany. News to them. Earl Trosino pays a visit to Dan Gallery at the Pentagon. Gallery has been promoted to the rank of rear admiral. He is now assistant chief of Naval Operations for Guided Missiles and has the clout to render assistance to Felix. He promises Trosino to see what he can do about making arrangements to bring Felix back to America. Gallery heads over to the State Department and meets with the head of the visa section. He signs statements citing Ewald Felix as an anti-Nazi hero who provided "invaluable assistance" in the capture of *U-505*. Thanks to Gallery, the wheels of government start churning.

Two weeks later, Trosino writes to Felix, giving him a heads-up that the US consul in Hamburg will be calling on him very soon.

"I know you will be pleased and see that I am trying to do all that is possible to get you here," Trosino writes.

"To start with there are three places for you to live and make your home. First with me, second with my mother, and third on Admiral Gallery's farm."

Unfortunately, Felix never receives the letter, and the US consul can't track him down in the bedlam. Felix is trapped in Germany, fearing for his life if he's ever outed as an American collaborator.

Several months pass before Felix's American friends hear from him again. Felix is now in Poland. He sends Trosino a heartbreaking plea—does the Navy commander have any secondhand clothes to spare for himself and his parents?

■ ■ ■

Harald Lange returns from the dead in May 1946, minus his right leg. And, yes, his wife, Carla, did remain faithful. They are living in Hamburg, raising their son and daughter. Lange finds work managing a fruit

import business at the harbor in Hamburg. His access to an abundant
supply of food makes him a good guy to know. But in the ranks of former
Kriegsmarine officers, there are those who believe Lange has a lot to
answer for. Even Gallery, back in America, hears rumors that Lange is
being ostracized by his Navy compatriots.

Axel Loewe, the first commander of *U-505*, has also fallen on hard
times.

After Germany's surrender, Loewe is briefly interned by the British.
Upon his release, he reunites with his wife, Helga, and their two teen-
aged children. All he possesses is the field gray naval uniform he's wearing. His parents had to flee the Loewe family estate in Mecklenburg
ahead of the Russians. The glory days are a distant memory. They have
lost everything, including the "floor from under their feet."

Loewe works on a street gang clearing debris and then as a farmhand
and manager of a sawmill on a large estate outside the seafaring town of
Flensburg in northern Germany. Food is scarce. Times are miserable. He
works from 6:00 a.m. until 6:00 p.m. for eighteen cents an hour. The
Loewes own a pig, which they fatten to four hundred pounds before
he is butchered. They also own chickens and grow potatoes and vegetables on a patch of land. Helga gives birth in 1948 to another daughter.
Loewe feels lucky he still has his health, and not much more.

He's still devoted to his former admiral, Karl Doenitz, now serving a
ten-year sentence in Spandau Prison in West Berlin following his conviction at the Nuremberg war crimes trials.

"A truly great man, and to this day I stand by him," Loewe says.

EPILOGUE

THE QUIET MEN

Earl Trosino can't believe this is really happening. Ten years have passed since the capture of *U-505* and Trosino now finds himself back on the submarine, this time as her skipper.

U-505 is making her way out of the Portsmouth, New Hampshire, Submarine Naval Yard. A tugboat is towing her up the St. Lawrence River, through the Welland Canal in Canada that connects Lake Ontario to Lake Erie, then it's on to Lake Huron and finally Lake Michigan. Four of the five Great Lakes must be traversed before *U-505* reaches her final destination—the Museum of Science and Industry in Chicago.

Trosino finds *U-505* to be sound, seaworthy, and watertight, but still a beast—"evil in her looks and a problem to handle." The boat reminds Trosino of a vampire—"out for your blood."

He is as terrified of the submarine on this voyage as he was when he salvaged her on the high seas in 1944. In his estimation, she is a malignant instrument of war, like all *unterseebooten*.

So, how in the world does Trosino, of all people, end up as the last skipper of *U-505* in the spring of 1954? As Dan Gallery would say, it's a hell of a story.

At the end of the war bond drive in 1945, the Pentagon, having no further use for *U-505*, sailed her to the Navy base in Portsmouth. There, the submarine was left in limbo, neither decommissioned nor put in mothballs.

What to do with *U-505*? Sink her in deep water, like the other U-boats from the Nazi fleet? Scrap her for steel? When Dan Gallery heard those ideas, he raised a fuss and used his clout to make sure that *U-505* would never be scuttled. So, she just sat there, exposed to the elements and salt water, rusting and forgotten, cannibalized over the years of all instruments, piping, furnishings, brass fittings, and vital installations. The radio room and sound room became empty shells. The crews' quarters and control room were encrusted with salt and rust. The bunks were thrown on top of one another into a pile of junk. The huge diesel engines were inoperable. Nothing really worked. The deck was also rotting, and the gangplank raised because the Navy deemed the sub too dangerous to board.

It was Gallery's brother, Father John Ireland Gallery, who came up with the idea of transporting *U-505* to the Gallery family's hometown of Chicago. What a splendid permanent exhibit she would make for the Museum of Science and Industry, as a memorial to the fifty-five thousand sailors and merchant marines who were lost at sea during the Battles of the Atlantic in the two world wars of the twentieth century.

It took years of lobbying. Colonel Robert McCormick, the influential publisher of the *Chicago Tribune*, threw his newspaper's editorial weight behind the project. Negotiations ensued with the Navy to obtain legal title, which required an Act of Congress. Funds had to be raised to make *U-505* seaworthy and pay for the expense of transporting her a thousand miles inland to Chicago. The chipped paint and rust and mussels clinging to the outer hull had to be sandblasted away and the superstructure spruced up and given a new coat of paint. Regrettably, the paint selected was black, which was the color of US Navy submarines. This blunder was later put right and *U-505*'s original color scheme restored for historical accuracy to a two-toned gray exterior.

When everything finally came together, Gallery reached out to Trosino: Would he be interested in commanding *U-505* on her last journey?

"Are you kidding?" Trosino wrote back. "You made me responsible for 'Junior' way back when and I certainly would like to accompany her to her final resting place."

Gallery responded, "Nothing would be finer, in my opinion, than to have you bring *U-505* to Chicago. So it's a deal."

Gallery arranged for Sun Oil to grant Trosino a leave of absence from his current assignment as chief engineer on the oil tanker *Maryland Sun*. Trosino takes the helm for a five-week journey that ends in June, when *U-505* lands at the Michigan Avenue Bridge in the Windy City.

The next monumental engineering task is jacking the submarine six feet high and transporting her across one of the busiest thoroughfares in America—Lake Shore Drive. It's like raising a house from its foundation and relocating it to the other end of the street, except this monster weighs 850 tons. On the evening of September 3, after the evening rush hour, the drive is closed to traffic and *U-505* is wheeled across the highway on a steel cradle at the rate of eight inches a minute as a crowd of fifteen thousand spectators watches and cheers.

The following dawn, Lake Shore Drive reopens to traffic in time for the morning rush.

U-505 is now nestled on three huge concrete cradles in the backyard of the science museum. The official dedication ceremony is held on September 25, 1954, before a crowd of forty thousand people and broadcast live on national television. The master of ceremonies is TV and radio star Arthur Godfrey. Fleet Admiral William "Bull" Halsey delivers the keynote address. Rear Admiral Dan Gallery speaks about the submarine as a fitting memorial to the sailors who lost their lives in war, as there are no tombstones at sea. He also introduces the members of the boarding party. They're all there: Mack Pickels, Zeke Lukosius, Chester Mocarski, Gordon Hohne, George Jacobson, Stanley Wdowiak, Arthur Knispel, William Riendeau. All except Lieutenant David, dead now nine years. Earl Trosino is also present and given the honor of unveiling the bronze plaque that depicts *U-505*'s capture.

A choir sings "Eternal Father, Strong to Save," the Navy Hymn. A Marine honor guard fires a volley. A bugler stands on *U-505*'s conning tower and plays "Taps."

Gallery has a great time reconnecting with his boys. They may have gotten a little thicker around the waist and thinned out on top, but they remain the same high-spirited rascals who went beyond the call of duty all those years ago. Of course, there's still the thorny matter regarding the spoils of war. Yep, they're still griping, and Gallery keeps hearing, "Cap'n, where the hell is that Luger you made me turn in and were going to get back for me?"

Over the following years, U-505 came to be restored with fittings and spare parts from other U-boats. The German company that manufactured the original diesel engines also contributed engine components. In 2004, after a half century of outdoor display in the harsh Chicago climate, U-505 would be moved once again to be preserved for future generations. Her new home is four stories underground, in a forty-two-foot-deep, temperature-controlled pen at the Museum of Science and Industry, where she remains on exhibit to this day.

There are those who maintain the capture of U-505 shortened the duration of the war, perhaps by several months. Even Gallery—a master self-promoter—was skeptical about making such an assertion.

It makes for a lively debate. Imagine the mayhem to America and Great Britain if three hundred U-boats had not been sunk in the wake of U-505. Would the fanatical Nazis in the Kriegsmarine have gone on to fight another day for the Third Reich? How many Allied ships and merchant marine vessels and lives were spared due to the events of June 4, 1944? It's one of the great "what-ifs" of World War II.

■ ■ ■

Remember how hard it was for everyone involved to keep mum about U-505 until the war was won? Once they came home, a funny thing happened—they really didn't like talking about it.

Chester Mocarski never told his wife how he nearly lost his life when he fell between the whaleboat and U-505. Not a word. She only learned what happened when Dan Gallery wrote a letter to Mocarski inviting him to attend a fundraising event in Cleveland. What's U-505? she asked her husband. Only then did he tell her, and it was like pulling teeth.

Gordon Hohne's wife, Norma, also didn't have a clue until 1953 when a letter arrived at their home in Massachusetts. She saw the official US Navy letterhead and thought, dear Lord, her husband was being called to active duty to fight in the Korean War. When they opened the letter, she saw it was from Dan Gallery, asking Hohne to send a photograph of himself for the unveiling in Chicago.

"Guess we're going to Chicago," Hohne said.

"Why?"

"Well, we captured this German U-boat . . ."

Flags Hohnes had three children, Robert, Wendy, and Pamela. One day, when young Robert Hohne was home on summer break from college, his mother retrieved a black case from the bedroom nightstand and showed it to her son. Inside was a Silver Star.

"Your dad was a war hero," she told him.

In old age, Hohne was invited by his grandson Matthew's third-grade teacher to give a speech to the class about his wartime adventures. Matthew had to beg his "Papa" to do it.

Hohne didn't bring his Silver Star. No theatrics for this old salt. He just pulled up a chair to the front of the classroom, sat down, and spun his story.

Hohne stayed in contact with the other guys from the *U-505* boarding party. He and Wayne Pickels exchanged Christmas cards and birthday greetings. They spoke once a month or so and kidded each other about where they lived, with the conversation usually going something like this:

Pickels: "How's the snow in Massachusetts?"

Hohne: "How's the heat in Texas?"

Every ten years, as long as health permitted, they attended a reunion of the boarding party at the Museum of Science and Industry. They watched one another grow into old men.

After his retirement from Otis Elevator, Hohne kept busy. He played in a candlepin bowling league and enjoyed billiards and cribbage and watching the Boston Celtics and the Red Sox on TV. He was diagnosed with prostate cancer. His doctor told him he needed a heart valve replacement.

"What are my chances of making it off the table?" he asked.

Maybe 20 percent.

He thought about it. "Naw. I've lived a good long life. A lot of my brothers didn't make it back."

The flag man died at age ninety. And that American flag switcheroo he pulled off way back in 1944 on the USS *Pillsbury*? After all these years, Old Glory was still hanging in a cabinet in his finished basement in West Boylston, Massachusetts.

■ ■ ■

Mack Pickels worked in customer service for Westinghouse and rose up the ranks to manager. He liked telling the story of an ornery customer who complained nonstop about the loud noise coming from her refrigerator. When he got to her house and saw there was absolutely nothing wrong, Mack stood before the fridge, waved his hands like a magic wand, and pronounced, "Abracadabra!" As far as the customer was concerned, that spell did the trick.

Pickels's best friend from the boarding party was Zeke Lukosius. They regularly attended *U-505* reunions in Chicago where, on big anniversaries, they would run into former German sailors who were making their own pilgrimages to the science museum with their wives to show them the U-boat that had changed their lives.

Time had worn away the grudges.

"Yesterday's enemies are today's friends," Pickels liked to say.

Sometimes, however, the encounters would get a little edgy. There was one long-retired German sailor who pointed at the submarine with pride.

"You know, that's my boat," he said.

"It *was* your boat," Pickels responded. "But we took it from you."

Joseph Villanella, the radarman from the USS *Chatelain*, was attending a *U-505* reunion in 1999 when he ran into the German sailor Werner Reh.

Of all people, Reh!

Villanella had to chuckle. That was a name he could never forget. He had found Reh's underwear laid out on the deck of the *Chatelain* way back on June 4, 1944. This was after the Germans had been issued clean G. I. dungarees. The underwear was Villanella's only souvenir from that eventful day, and all these years later, he still possessed it, back home in

Hudson, New Jersey, where he ran a laundry business. He had once vowed to return the underwear to Reh one day in the unlikely event their paths should cross again.

Now, out of nowhere, here was Werner Reh. Unbelievable. Villanella struck up a conversation. Reh didn't speak English very well, but with hand gestures and a few choice words, Villanella conveyed to Reh that he still possessed his underwear—and he'd be happy to return it if Reh would supply his address in Germany.

"They won't fit him anyway," Villanella remarked. Alternatively, he suggested hanging the underwear on the *U-505* mast.

Well, not all of yesterday's enemies are today's friends. For some Americans, Germany's instigation of a world war was never to be forgotten or forgiven.

■ ■ ■

When he retired from Westinghouse, Pickels bought an RV. Once a year he and his wife drove to Chicago to see *U-505* and his buddy Zeke Lukosius, who lived outside the city.

"It shaped my dad's life in a lot of ways," says Pickels's son Douglas. "He loved his family and his country. That's why they're called the Greatest Generation."

Wayne "Mack" Pickels died at the age of ninety-one, the last surviving member of the boarding party. Douglas was handed the America flag that draped the coffin. Pickels's other son, David, got his Silver Star.

■ ■ ■

Lukosius died in 2006 at the age of eighty-seven.

"He never embellished the story or his part in the capture," said his daughter, Diane. "It never turned into a fish story with him. If you heard him tell it once and then heard him tell it again twenty years later, it was always the same." To the end, he maintained he wasn't a hero, just a regular sailor doing his job.

■ ■ ■

Stanley "Stash" Wdowiak didn't want to talk about his war experiences either. In fact, his wife, Sissy, never connected her husband to *U-505* until she read a magazine article about the capture in 1954. There was his name!

Over time, Stash loosened up. He kept his Navy Cross in a glass box in his son Peter's bedroom in Queens. He also held on to that pair of Zeiss binoculars that he had somehow managed to smuggle out. Stash would take the binoculars with him to New York Mets games to watch the action on the field from the nosebleed section of Shea Stadium. It was quite the conversation piece when fans sitting next to him took note of the beaten and bent swastika and Nazi eagle engravings. Stash got a kick out of relating the backstory to those who wondered what he was doing with Nazi memorabilia, in New York City, no less. He sure set them straight.

All his grandchildren have made the pilgrimage to the *U-505* exhibit in Chicago. Stash's Navy Cross and his Zeiss binoculars are now on exhibit at the museum.

He died at the age of sixty-three, the aftereffects of a debilitating stroke. Today, the street where he lived is named after him.

■ ■ ■

Earl Trosino retired from the Navy Reserves in 1964 with the rank of rear admiral. He and Lucy adopted two orphaned brothers from Italy, Gerry and Joseph, and lived in a colonial brick house with a one-car garage on Sunnybrook Road in Springfield, Pennsylvania, which the Trosinos had purchased in 1939 for a thousand dollars. It stood on the corner lot where they could see everything going on in the neighborhood in all directions. Nothing fancy, but a sweet, sweet home.

In 1989, he paid a visit to the Museum of Science and Industry and, much to his surprise, bumped into Hans Doebler, a former U-505 crewman who had immigrated to the United States seven years earlier. They reminisced. Doebler was sixty-five, short, barrel-chested, and muscular.

When no one was looking, Doebler bounded his way up the *U-505* exhibition ramp.

"We can't go up there!" Trosino called out. "You put the museum in legal jeopardy if something happens."

Behaving like a disobedient adolescent, Doebler ignored Trosino. Though Doebler had eighteen years on the retired rear admiral, no way was Trosino going to let the German get the better of him. He strode up the ramp.

Moments later, these two graying sailors stood together on the conning tower. Memories came flooding back to Trosino, to that day in 1944.

"I had to grab the cable and work my way up here hand over hand, with the sea banging me right and left. When I got to the conning tower I didn't have the strength to pull myself up. But the signalman from the *Pillsbury* [that would be "Flags" Hohne] reached down for me, and that's how I managed to climb up here."

He looked at Doebler.

"In 1944 we tried to kill one another." Now, he reflected, "We're like brothers."

In 2001, Earl Trosino was sent to a nursing home to live out his final years. When his son Joe came to visit, the old warrior told the orphan he'd rescued from Italy all those years ago, "Joe, get me the hell out of here." He lasted just three weeks. He died from prostate cancer at the age of ninety-six.

■ ■ ■

Rear Admiral Dan Gallery decided the time had come to write a book about *U-505*. To that end, he engaged in a lively correspondence in 1955 with Axel-Olaf Loewe in Germany, seeking information about Loewe's career and the early days of *U-505*.

"As you may have heard," he wrote Loewe, "*U-505* was eventually captured by an aircraft carrier task group under my command and is now in Chicago at the Museum of Science and Industry. After you left the *U-505*, what did you do? What are you doing now? Has there been any publicity in Germany about this capture, and if so what was the opinion?"

Loewe wrote back, "My most honoured [sic] Admiral!" He promised his full cooperation. "The loss of this boat was never known in Germany, and as far as I know, not even to the High Command," Loewe wrote. He said the seizure of *U-505* only received widespread attention in West Germany in 1954 when she sailed to Chicago. The letter ended, "With highest esteem, Your very devoted, Axel Loewe."

The Museum of Science and Industry invited Loewe to Chicago to view his old boat, but Loewe said his family had "lost their whole fortune" and he could not afford the overseas trip unless he was paid a stipend of $130 a week.

In 1957, Loewe made it to Chicago and was given a VIP tour of *U-505*. He died in 1984 at age seventy-five.

Harald Lange also came to Chicago, in 1964, on the twentieth anniversary of the boat's seizure. Gallery handed him a memorable souvenir—Lange's own Zeiss binoculars, which he had last possessed on June 4, 1944.

Lange did not live long after that. He returned to Germany and died at age sixty-three in 1967.

Hans Goebeler worked as a welder and tool-and-die maker in Germany, and after his retirement moved with his wife and daughter to the land of his former enemy, the United States, which he now viewed as the bulwark against the postwar communist threat in Europe. He chose to live in a Chicago suburb to be close to his precious boat. Later, he moved to Florida for his wife's health and died of cancer in 1999 at age seventy-five.

Another German *U-505* crew member also couldn't stay away from the seduction of the boat. Diesel Machinist Hans Decker lived in Hamburg but expressed a desire to become an American citizen and pulled up stakes, immigrating to Chicago with his wife and two children. Museum officials were so impressed with his knowledge of the U-boat's equipment, they offered him a job as the exhibit's maintenance engineer. Occasionally, he also conducted tours of *U-505* for visitors. Gallery took a liking to Decker. When they ran into each other at the museum, Gallery put his arm around him and said, "I would have been proud to have had you as a crew member of my ship."

Decker replied, "I am sure, sir, that everybody now agrees that war is brutal nonsense."

Dan Gallery was mentored in his new authorial profession by Pulitzer Prize–winning historical novelist Herman Wouk. Gallery was an admirer of Wouk's breakout success, *The Caine Mutiny*, which Navy brass detested because of its depiction of the unstable fictional character Captain Queeg. But Gallery loved the book and wrote Wouk a fan letter in which he declared he "couldn't put it down—and sat up till 5 a.m. this morning to finish it." Gallery invited Wouk to meet him at the Oak Room in Manhattan for a drink. Wouk was first to arrive. In walked Gallery. He was, in Wouk's words, "A short lean man in a brown jacket and a bow tie, with large ears and a quizzical face."

"Herman Wouk? I'm Dan Gallery."

Gallery got to the point. He wanted to know if Wouk could recommend a literary agent. He had been hit with the writer's bug when he wrote that *Saturday Evening Post* article in 1945 and now Gallery wanted to pursue a second career as a writer.

Gallery went on to write five works of nonfiction and five novels, including *The Brink*, a Cold War thriller about a nuclear submarine named *Nemo*.

He died at the Bethesda Naval Medical Center from emphysema at age seventy-five. His brothers, William Gallery and Phil Gallery, now rest next to him at Arlington National Cemetery. All three Gallery brothers achieved the rank of rear admiral, a first in US history.

Herman Wouk, amazingly, lived to the age of 103. He was truly an old salt, having served as an officer on minesweepers in the Pacific as Gallery's men were carrying out their miracle at sea.

■ ■ ■

And what about Albert David, who died just nineteen days before he could savor the accolades of a White House ceremony from the president of the United States? In 1964, the Navy announced the name of its newest destroyer escort, the USS *Albert David*, in honor of the officer in charge of the boarding party that captured *U-505*. A routine press release was sent to the *Daily Forum* newspaper in Maryville, Missouri, where, according to Navy records, Albert David had grown up. Naming a US warship after a local hero is a big deal for a small town like Maryville. But who was Albert David? Incredibly, nobody in Maryville seemed to remember.

The editor of the newspaper expressed his frustration in a front-page column.

"This has been one of those days when things just didn't seem to gel. Much time was consumed today in trying to learn the identification of an Albert David, listed as an ex-enlisted man from Maryville. Numerous telephone calls have been made to various individuals whom we thought could give us some information, but to no avail. So now we're asking you readers: 'Do you know Albert David?' He got the Congressional Medal of Honor, the only one awarded in the Battle of the Atlantic.

"We have searched hither and yon, but all we come up with is a blank where the name of Albert David is concerned. Again we ask, 'Do you know Albert David?' If anyone can give us a clue as to his identification a call to the *Forum* office would be much appreciated."

Six days later, the editor gave up.

"Nobody here seems to know the hero." The editor's conclusion? The Navy had obviously messed up. "He must have been from another Maryville or Maysville and we got the credit mistakenly."

Maybe, he suggested, this obscure lieutenant came from Merryville, Louisiana. Or Maryvale, Utah. Or Maysville, Kentucky. But definitely not here, in Maryville, Missouri.

Well, newspaper editors are grouchy when they don't get what they want, but here's what really matters about Albert Leroy David. He left home at age seventeen, in 1919, and went to sea for a long, honorable career. He was awarded the Medal of Honor for climbing down a pitch-black hatch and boarding a Nazi submarine knowing he could be cut down by a hail of bullets and fully willing to take that chance if it led to the success of the mission.

Where he came from doesn't matter. Where he went in the name of liberty is what counts.

ACKNOWLEDGMENTS

This book could not have been written without the support of the Museum of Science and Industry in Chicago. I want to especially thank Kathleen McCarthy, the director of collections and head curator, and collections and archives volunteer Doug Zimmer, who were indispensable in guiding me through the vast *U-505* collection, which includes oral histories, letters, and important government documents concerning this remarkable episode in World War II history.

I am deeply grateful to my literary agent, Frank Weimann of Folio Literary Management, in taking on this project, and to my friend Lisa Sharkey. Keith Wallman, editor-in-chief at Diversion Books, will forever have my appreciation for acquiring the book.

The extraordinary author Charlie Carillo read the manuscript and offered incisive suggestions that greatly enhanced the final product. I want to thank my wife, Nancy Glass, who always supports and encourages my work. Her comments on the manuscript contributed significantly to the editing process.

Rear Admiral Daniel Gallery wrote lively accounts of his life in the US Navy. The two that were most useful in my research were *Twenty Million Tons Under the Sea* and *Clear the Decks!* The German submariner Hans Goebeler wrote about his experiences as a crew member on *U-505* in *Steel Boat, Iron Hearts*, coauthored with John Vanzo. Anyone contemplating a book on *U-505* must acknowledge Goebeler's vivid portrayal of what it was like to fight on board a U-boat during the war, as depicted in *Steel Boat, Iron Hearts*.

I have attempted to tell the story of *U-505* from the point of view of the nine Americans who were members of the boarding party that seized the submarine on June 4, 1944. They are all gone now, but in 1999 several of them were interviewed at the museum for a *U-505* documentary, and they left behind a remarkable oral history that I have relied on to write *Codename Nemo*. A number of German sailors who served on *U-505* also participated in the documentary. Their names are referenced in the footnotes.

The sons, daughters, and grandchildren of the USS *Pillsbury* boarding party were generous with their memories.

Thanks to Wayne Pickels's son Douglas and grandson Daniel for their family remembrances and giving me access to Pickels's wartime diary.

Thanks also to Robert Hohne, Pamela Genelli, Nicholas Genelli, Matthew Genelli, Ed Bednarczyk, Nick Fascia, Patricia Mascia, and Tom Pinamonti.

I am indebted to Joseph and Janet Trosino for permitting me access to Trosino family documents as I researched the life of the remarkable Earl Trosino.

Thanks also to the works of Joseph Scafetta Jr. and the historian and U-boat expert Eric Wiberg, and the Museum of Science and Industry's former *U-505* curator, Keith Gill.

For my research in Bermuda, thanks to Paget Dackman, Andrew Birmingham, Edward Harris, and the National Museum of Bermuda's registrar, Jane Downing.

Appreciation is also extended to Henry Lachman for his considerable research skills, and to the Wisconsin Veterans Museum, the Wisconsin History, the Nimitz Library at the US Naval Academy, the US Naval Institute, and the Pritzker Military Museum & Library.

Bill O'Reilly was very kind to read an advance copy of *Codename Nemo*.

Nate Becker introduced me to the glories of the *U-505* exhibition at the museum in Chicago. I urge everyone to see it at least once.

Any errors of fact or omission in *Codename Nemo* are mine alone.

As always, my most important acknowledgments go to my wife for her loving support and guidance; to our children, Max, Pamela, and Sloane; to Alice Heath; and to the newest member of our family, Arthur I. Lachman-Heath. When he's old enough, may he one day read *Codename Nemo* and contemplate the sacrifices and courage of the generation of men and women who fought in World War II and saved the world from tyranny.

SOURCES

Prologue: June 29, 1942

xviii "How hard are we hit?" McCarthy asks: Account of the sinking of the Thomas McKean based on statements by Chief Engineer Thomas McCarthy to the Maritime Administration, US Dept. of Commerce, June 26, 1942.

Chapter 1: "Where the Hell Is Freetown?"

1 Loewe comes from a seafaring family: Axel-Olaf Loewe letter to Daniel Gallery, October 28, 1955, Museum of Science & Industry (MSI).

2 "Comrades, as commandant of U-505": Loewe quoted in *404 Days! The War Patrol of the German* U-505 by Hans Joachim Decker (March 1960, US Naval Institute).

2 Second in command is the chief engineer, Fritz Forster: Theodore P. Savas, editor, *Hunt and Kill: U-505 and the U-boat War in the Atlantic* (Savas Beatie LLC, 2012), 28.

4 "Comrades, the Fatherland is now behind us and our first patrol is underway": Decker, 34.

5 "*U-505* reporting as ordered to the Second U-Flotilla!": Decker, 34.

5 Storing food for fifty men to last one hundred days at sea: Loewe to Gallery letter, MSI.

6 In all, U-505 will set sail with twelve tons of food: Loewe to Gallery letter, MSI.

7 Hans Goebeler reports for duty: Hans Goebeler with John Vanzo, *Steel Boat, Iron Hearts*: A U-boat Crewman's Life Aboard *U-505* (El Dorado Hills, CA: Savas Beatie, 2008), 13.

8 "One small mistake can sink a boat": Goebeler, *Steel Boat, Iron Hearts* 18–19.

8 *Leinen los!*—"Cast off!": Decker, 35.

Chapter 2: This Cold Strange Land

10 They "vant to be alone": Daniel V. Gallery (New York: Paperback Library, 1970), 12.

11 Gallery graduated Annapolis in the top 10 percent of the class of 1921: C. Herbert Gilliland & Robert Shenk, *Admiral Dan Gallery: The Life and Wit of a Navy Original* (Annapolis, MD: Naval Institute Press, 1999).

11 "I think I'm going to get married . . . to Miss Vera Lee Dunn: Gilliland, *Admiral Dan Gallery*, 21.

11 Miss Dunn's "lack of social status": Gilliland, *Admiral Dan Gallery*, 64.

11 "I wouldn't have the crown princess of England if I could have Vee instead": Gilliland, *Admiral Dan Gallery*, 69.

12 Relief map of the Emerald Isle: Daniel V. Gallery: *Twenty Million Tons Under the Sea: The Daring Capture of the U-505* (Washington, DC: Regency Publishing, 1956), 63.

13 Fact is, they see eye to eye: Gallery, *Clear the Decks!* (New York: Paperback Library, 1967), 13.

13 Gallery orders Navy Seabees to construct new barracks: Gallery, *Clear the Decks!*, 28.

14 "kwitcherbellyakin": Gallery, *Clear the Decks!*, 31.

17 Gallery gets to thinking: Gallery, *Clear the Decks!*, 42.

Chapter 3: Kill! Kill! Kill!

18 *Aufbacken!* "Chow's up!" Strong coffee is served: Goebeler, 23.

20 *Tauchen!* Loewe calls out. "Alarm! Dive! Dive!": The sinking of the Benmohr is based on the writings of Decker, Goebeler, and *Hunt and Kill*, 63.

25 Under international laws of war: Jon L. Jacobson, *The American Journal of International Law*, Vol. 87, Chapter VIII, "The Law of Submarine Warfare Today," 205–206.

25 "sank in 25 seconds": Decker, 26.

26 A sailor dresses as Neptune in a flowing white beard and robe: Goebeler, 29.

27 Loewe's "nerves of steel": account of the sinking of the West Irmo: Goebeler, 30–31.

27 No point in sticking around and picking a fight: Gallery, *Twenty Million Tons*, 91.

Chapter 4: The Eyes of Texas Are Upon You

29 Wayne Jr.'s upbringing was as wholesome and all-American: Harry B. Orem, editor, Wayne Macveigh Pickels, BM 1/c, United States Navy, Silver Star Recipient (San Antonio, Texas) privately printed, 1999.

30 "Howz That?": *San Antonio Light*, May 27 1942.

31 "We are proud to welcome you into the Navy": Associated Press, June 2, 1942.

31 "Have faith in the Lord.": Pickels family documents. Courtesy of Pickels Family.

33 "We had a big laugh over that": Pickels war diary, November 29, 1943.

33 Two palm trees, which he plants at the entrance to the base: Gilliland and Shenk, *Admiral Dan Gallery*, 93.

33 "FBI," as in "Forgotten Bastard of Iceland": Gallery, *Clear the Decks!*, 46.

Chapter 5: Sin City

36 Loewe's mother is of Dutch descent: Loewe to Gallery, October 18, 1955, MSI.
36 The ship's master is Reindert Johannes van der Laan: U-boat.net: SS *Alphacca*.
37 "Beginning today we start our trip home": Decker, 37.
38 His mother is a devout Christian. Each letter opens with a verse from the Bible: Goebeler, 190.
39 A pretty French girl who calls herself Jeanette: Goebeler, 94.
40 "First mission of captain with new boat, well and thoughtfully carried out": Goebeler, 42.
40 "Men, we've been ordered to the Caribbean": Decker, 38.
41 Hans Goebeler is on bridge watch duty as *U-505*: Goebeler, 47.
42 By some miracle, there are no casualties: u-boat.net: SS *Sea Thrush*.

Chapter 6: Sitting Duck

43 Back in his glory days as a champion all-American swimmer, Albert Rust Jr.: Greensburg, Indiana: *Daily News*, February 8, 2013.
44 And this is where Rust finds himself on April 28, 1942: Albert Rust Jr., oral history, March 3, 1999, MSI, 20.
47 Lifeboat #3—commanded by Second Mate Roland Foster: statement from Foster, Maritime Administration, US Dept. of Commerce, June 26, 1942.

Chapter 7: Relieved of Command

49 "Fire a shot over her bow so we can find out who she is": Account of the sinking of the Roamar based on Goebeler, 53–59; Decker, 39; *Hunt and Kill*, 20.
50 Nothing can be accomplished unless the commandant is in fit condition: Loewe to Gallery, November 15, 1955, MSI.
51 "a little hollow": Goebeler, 59.
51 The new skipper of *U-505* is Kapitänleutnant Peter Zschech: *Hunt and Kill*, 31.
51 but in Loewe's opinion he has a "rather weak constitution": Loewe to Gallery, November 15, 1955, MSI.
52 "A thoroughly unpleasant character": Goebeler, 71.
52 In reality—at least in Goebeler's opinion—he knows "almost nothing": Goebeler, 62.
52 Even Thilo Bode agrees Hauser is "not an easy man": Thilo Bode, oral history, MSI.
53 "Kapitänleutnant Loewe is no longer in command of this boat!": Goebeler, 66.
54 "Maschinengefreiter Goebeler here as ordered, Sir!": Goebeler, 72+.
55 The roar of the diesels tells him the chase is on: Goebeler, 75.

Chapter 8: "I Deeply Regret to Inform You . . ."

57 Flight Sergeant Ron Sillcock is the ace pilot of Squadron #53 of the Royal Australian Air Force: Ken Sillcock, *Two Journeys into Peril: Wartime Letters of Ken & Rain Sillcock, 1940–1945* (Melbourne, Australia: Clouds of Magellen Press, 2009).

58 Does he mistake it for a shimmering bird in the tropical sky?: Decker, 40.

58 They see Lieutenant Stolzenburg sprawled unconscious on the deck: Goebeler, 84. The account of Sillcock's attack mode is based on *The U-boat War in the Caribbean*, by Gayklor Kelshall, quoted by Goebeler, 87.

59 It's a decapitated head. A blond head: Bode, MSI, 13.

59 "You can do what you want, but the technical crew is staying": Bode, MSI, 84.

60 banging away with sledgehammers: Bode, MSI, 90.

61 "Glad you made it home!": Bode, MSI, 105.

61 "I deeply regret to inform you that no trace": Sillcock, 277.

Chapter 9: Call to Duty

63 Earl Trosino papers, courtesy of Joseph and Janet Trosino.

Chapter 10: Sabotage

66 A delighted Goebeler steps into his favorite whorehouse: Goebeler, 113.

66 Radio operator Karl Springer heads to his village: Oral History Project, March 1, 1999, MSI.

67 Torpedoman Wolfgang Schiller really lives it up on his leave: Wolfgang Schiller, Oral History Project, March 1, 1999, MSI.

67 He is "proud as a peacock": Goebeler, 126.

68 sausages and smoked meats and baked goods: Goebeler, 126.

68 scrutinizing the "tasty dishes" of Lorient: Goebeler, 150.

70 Zschech's new second in command, or first watch officer, is twenty-six-year-old Paul Meyer: *Hunt and Kill*, 36.

70 But there's trouble right away: Goebeler's account of sabotage, 138+.

70 "Quick!" he cries. "Take us down to sixty meters.": Decker, 42.

71 "Holy shit!": Goebeler, 143.

72 "It's not true!" Springer shouts back: Springer, oral history, MSI.

73 This time it's a lone musician playing the harmonica: Decker, 43.

Chapter 11: USS *Can Do*

74 Captain Dan Gallery is hardly bowled over when he sees the USS *Guadalcanal*: Gallery, *Clear the Decks!*, 47+.

75 maybe the Saratoga: Gilliland & Shenk, *Admiral Dan Gallery*, 101.

75 Henceforth she is known as the USS *Can Do*: Gallery, *Clear the Decks!*, 50.

75 "the most primitive I've ever seen on a ship that size": Gilliland & Shenk, *Admiral Dan Gallery*, 98.

76 "Cap'n, I'd swap 'em all": Gallery, *Clear the Decks!*, 58.

76 "cracks of doom": Gallery, *Clear the Decks!*, 68.

77 "This is a hell ship," Trosino blurts out: Earl Trosino, Oral History Project, March 3, 1999. MSI.

80 Lieutenant Berg arranges for a bus to take Gallery: Naval History and Heritage Command. *Guadalcanal* Memory Log, 1943–1958.

81 When the order comes through, Gallery's heart sinks: Gilliland & Shenk, *Admiral Dan Gallery*, 98.

Chapter 12: Over the Side

82 A "cold clammy coffin": Goebeler, 165.

84 The pressure hull rings "like a church bell": Goebeler, 169.

85 "So, please, doctor, be quiet": Goebeler, 171.

85 Radio operator Karl Springer runs into the chief machinist: Springer, oral history, MSI.

86 "First Officer speaking. The captain is dead": Decker, 43.

Chapter 13: Turkeys in the Air

88 This morning on the flight deck I counted 10 officers: Document from the personal collection of former USS *Guadalcanal* officer Elton N. Thompson.

89 Gallery wears his cap at a jaunty, nonregulation angle: David Kohnen: "Tombstone of Victory: Tracking U-505 from German Commercial Raider to American War Memorial 1944 to 1954," *Journal of America's Military Past* (Winter 2007).

89 "When the going gets tough, they always send for the sons of bitches": Gallery, *Clear the Decks!*, 92.

90 New Year's Eve 1943: *New York Times,* January 1, 1944.

90 "there will be no victors": *New York Times,* January 1, 1944.

91 "now get going": Gallery, *Clear the Decks!*, 5.

91 ". . . 90 percent bad luck," *Clear the Decks!*, 160.

91 January 5, 1944: First day out is rougher: Pickels's war diary.

93 This is no time for rubbernecking: Gallery, *Twenty Million Tons*, 203.

93 landing on a baby flattop at night is maybe three times more difficult: Gallery lays out the difficulty of landing on a baby flattop in *Twenty Million Tons*, 202.

94 "The hell with it," he barks. "Shove it overboard.": Gallery, *Twenty Million Tons*, 202.

94 "That tail doesn't stick out very far": Gallery, *Twenty Million Tons*, 203.

96 "Everyone is sleeping in dungarees with a lifebelt on tonite.": Pickels's war diary, January 20, 1944.

Chapter 14: A Fragile Character

97 Goebeler passes, he gets the jitters: Goebeler, 178.
98 "It must be quite a surprise to you that your friend Peter": Goebeler, 181.
98 Zschech's bride must be informed: Thilo Bode oral history, MSI.
99 He always considered Zschech to be a fragile character: Loewe to Gallery, September 29, 1955, MSI.
99 Loewe concludes that Zschech was "consistently pursued": Loewe to Gallery, September 29, 1955, MSI.

Chapter 15: The Old Man

101 Kapitänleutnant Harald Lange stands before the crew: *Hunt and Kill*, 139.
101 Lange is also a member of the Nazi Party, number 3,450,040: *Hunt and Kill*, 38–39.
102 Machinist Hans Decker had a "disillusioning leave": Decker, 44.
102 return to Lorient with horror stories: Goebeler, 190.
103 Here at last is a commander worthy of respect, Goebeler, 195.
104 "Damned Tommies!" Lange sputters: Goebeler, 195+
104 "and we are thankful that a kind fate": Gallery, *Twenty Million Tons*, 188.

Chapter 16: Lone Wolf

106 "Tubes One and Four, prepare for a surface firing!": Timothy P. Mulligan, *Lone Wolf: The Life and Death of U-boat Ace Werner Henke* (Westport, CT: Praeger, 1993), 117.
106 to "hasten her demise.": Hardy, *SS Ceramic*, 335.
106 REPORT AT ONCE: Mulligan, 119.
107 Munday is in the smoke room of the SS *Ceramic*: Hardy, 349.
107 "She actually resembles that legendary ship": Clare Hardy, *SS Ceramic: The Untold Story* (CreateSpace Independent Publishing Platform, 2012), 14.
109 8th Dec: I awoke feeling very sore: Munday diary, quoted by Hardy. 350+.
110 "once you get into Germany you won't get any food like this": Hardy, 355.
111 "Have you a son called Eric?": London *Standard*, February 18, 1943.
112 Henke stands at attention before Hitler: Mulligan, 127.

CHAPTER 17: Ping . . . Ping . . . Ping

113 Gould is a twenty-nine-year-old Annapolis grad: *Kitsap Sun* newspaper (Bremerton, Washington), March 28, 2001, 6.
113 "Cap'n," Gould says breathlessly. "I almost got him!": Gallery, *Twenty Million Tons*, 207.

115 A third of the planes on the *Guadalcanal* were wrecked: Gallery, *Clear the Decks!*, 174.

116 "All depth charges fell short," the pilot reports: Gallery, *Twenty Million Tons*, 209.

116 "tough customer who knows his business.": Gallery, *Twenty Million Tons*, 208.

117 On board the *Pillsbury*, Boatswain Mate Second Class Wayne Pickels: Pickels and Zenon Lukosius, oral history, March 3, 1999, MSI.

118 "Tell the skipper here comes one of those sons-of-bitches with ears!": Pickels's war diary, April 9, 1944. Author's note: Gallery never references *U-515* firing an acoustic torpedo. In fact, he says Henke "didn't shoot at me with either guns or torpedoes after surfacing . . ." The discrepancy between the accounts of Gallery and Pickels can only, at this point, be explained by the fog of war memories. However, it should be pointed out that Pickels wrote his diary entry the day of the encounter.

119 Headland leaves the bridge. He is curious to check out these POWs: Harvey Headland, oral history, March 3, 1999, MSI, 32.

120 "Captain, I have a protest to make!": Gallery's exchange with Henke is recounted in *Twenty Million Tons*, 209+. Author's note: Henke and the other *U-515* POWs arrived in Norfolk on April 26. Henke was separated from his crew and taken to a top-secret interrogation center at Fort Hunt in Virginia. On June 15, Henke was shot to death as he made a desperate escape attempt from an exercise enclosure after learning that he was to be sent to Canada, the first step in his eventual extradition to England, where he would face trial for his role in the sinking of the *Ceramic*. Many believe Henke wanted to be shot because he knew he probably faced hanging as a war criminal in England.

125 Hans Kastrup is the sole survivor: James Wise Jr., *Sole Survivors of the Sea* (New York: Naval Institute Press, 2013), 28.

125 The body is lowered into the sea for a fitting burial: Gallery, *Clear the Decks!*, 182.

126 He and Gallery kept in touch via postcards every Easter: Gilliland & Shenk, *Admiral Dan Gallery*, 113.

Chapter 18: Seriously, Dan?

128 "I've made up my mind": Reminisces of Captain Henri Smith-Hutton, US Naval Institute, August 1976, Vol 2, 402.

128 What if our boys were able to seize the U-boat before the Nazis "pull the plug": Gallery, *Clear the Decks!*, 196.

130 Knowles was born in Nebraska and graduated from the Naval Academy in 1927: *Baltimore Sun*, "Society in Anapolis: Miss Velma Sealy to Wed Lieut. Kenneth A. Knowles," February 28, 1936, 5.

131 F-21 has been tracking an unidentified U-boat since March 16: *Hunt and Kill*, 113.

Chapter 19: Radio Silent

133 Toni the cook brings them a pot of hot coffee mixed with rum: Goebeler, 212.

134 the boat is at this moment positioned at 44-39N 14-30W: *Hunt and Kill*, 113. Author's note: For more information on how *U-505* was tracked, *Hunt and Kill* is highly recommended.

Chapter 20: The Nine

135 "When and if we bring a sub to bay, we don't clobber it forthwith as we had with *U-515*": Gallery, *Twenty Million Tons*, 219. Author's note: The narrative is taken from Gallery's own words, which I have transposed into dialogue and the present tense.

136 "barnacles on the brain." It is so quiet, a "flake of falling dandruff would have made an audible noise": Gallery, *Twenty Million Tons*, 221.

137 "WILL BE HOT ON TRAIL TOMORROW X EACH ESCORT": Orem, editor, Wayne Macveigh Pickels, 33.

137 ship's chief engineering officer, Lieutenant Junior Grade F. M. Burdette. But then Casselman reconsiders: Orem, editor, Wayne Macveigh Pickels, 36.

138 "Why'd you volunteer for something like that?" he asks: Xenon Lukosius oral history, MSI, 10.

141 Hohne made his first stab at joining the Navy in April 1942 at age sixteen: Author interviews with Gordon Hohne's daughter, Pamela Genelli, July 24, 2022; and Hohne's son, Robert, August 16, 2022. Additional material: Gordon Hohne oral history, MSI.

142 BOARDING AND SALVAGE BILL FOR ENEMY SUBMARINES: Kohnen, 32.

144 "Do you think the Germans are just going to let us come aboard? Huh?": *Wisconsin State Journal* interview with Lieutenant Albert David, July 29, 1945, 4.

145 He must lug the thirty-foot-long, three-eighth-inch chain onto the U-boat: Wayne Pickels war diary, April 9, 1944.

Chapter 21: Just One More Night

146 Hans Goebeler has had enough of the finicky wonder weapons: Goebeler, 224.

147 A "fat morsel" of a ship, Lange writes in his war diary: *Hunt and Kill*, 130.

147 the diesel engines pumping away at a two-knot: Goebeler, 224.

148 "What a dismal trip," machinist Hans Decker bemoans: Decker, 44.

148 a whole bunch of false alarms: Gallery, *Twenty Million Tons*, 89.

149 everything is "hunky-dory.": Gallery, *Clear the Decks!*, 144.

150 A German U-boat just torpedoed the aircraft carrier USS *Block Island*: Gallery, *Clear the Decks!*, 199.

151 F-21 has been tracking the submarine since March 16: *Hunt and Kill*, 94+. Author's note: Knowles's information comes from an ingenious network of fifty-one shore-based High Frequency / Direction Finding stations—Huff/ Duff—which have been set up to intercept the German U-boat fleet's radio transmissions. For more, see *Hunt and Kill*.

152 Whoever that U-boat skipper is, he's a "very cautious": *Hunt and Kill*, 133.

152 he sees Earl Trosino approaching with a deeply distressed expression: Gallery, *Twenty Million Tons*, 229.

152 "Cap'n, we've got to quit fooling around here and get to Casablanca": Gallery, *Twenty Million Tons*.

153 Trosino pegs it at 200 rpm: Gallery, *Clear the Decks!*, 164.

153 That U-boat skipper's goose is cooked: Gallery, *Twenty Million Tons*, 230.

153 monkey wrench tucked into his pocket and a rag to wipe the grease: Trosino, oral history, 95. Gallery's generic description of ship engineers.

Chapter 22: Battle Stations

155 "They're in bad shape," says his machinist: Decker, 45.

156 "Battle stations!" Lange calls out: Decker, 45.

Chapter 23: "Away, All Boarding Parties!"

158 Right now he's positioned a hundred miles off Cape Blanco: *Hunt and Kill*, 131.

158 a "complete blank": Gallery, *Twenty Million Tons*, 231.

158 "You better pray hard at Mass this morning, Cap'n": Gallery, *Twenty Million Tons*, 230.

159 Reveille is sounded and the half-asleep crewmen roll out: *Scuttlebutt* (in-house newsletter for USS *Guadalcanal*, date unknown).

160 Now Ray Watts is on board the USS *Chatelain*: Ray Watts, Oral History Project, March 5, 1999, MSI.

162 "Frenchy to Bluejay—I have a possible sound contact": All radio transcripts from "Capture of the U-505 Sequence of Events from Action Reports, Ship Logs and Log of Communications," MSI.

162 "an old lady in the middle of a bar room brawl—she has no business being there": Gallery, *Twenty Million Tons*, 232.

162 "Left full rudder," Gallery barks. "Engines ahead full speed.": Action Reports, MSI.

163 "I put a shot right where he is," says Roberts from his Wildcat: Action Reports, MSI.

164 "Hey, you guys, there's a sub out there": Ritzdorf, oral history, March 5, 1999, MSI, 14.

164 "I'm gonna go to church every Sunday.": Joseph Villanella, Oral History Project, March 3, 1999, MSI, 2.

Chapter 24: Abandon Ship

181 A strange metallic banging or "clinking" is coming from above: Goebeler, 230.

182 "Take us up," Lange cries out. "Take us up before it's too late!": Decker, 45.

183 *Raus schnell!*—"Abandon ship!": Decker, 45.

183 Still wearing his black silk pajamas, Lange squints: Jack Dumford, Oral History Project, March 20, 1999, MSI.

183 Accounts of *U-505*'s last moments [before capture] based on writings of Goebeler, Decker, Wolfgang Schiller, and Karl Springer oral histories, and Harald Lange interview published in the German magazine *Kristall* (Hamburg, 1956).

184 Wolfgang Schiller, still clad in that purloined lambswool sweater: Schiller, oral history, March 1, 1999, MSI.

185 Goebeler tosses the sea strainer cover: Goebeler, 236.

Chapter 25: Hi-Yo, Silver!

187 "Well, fellows," he says, "we may not be back but we'll give them a hell of a fight": *Wisconsin State Journal*, July 29, 1945, 4.

188 "Save me, comrades," one German beseeches: Lukosius oral history, March 3, 1999, 14, MSI.

189 "Hi-yo, Silver—ride 'em cowboy!": Gallery, *Twenty Million Tons*, 234.

189 "Roger. Nice going.": Action Reports, Ship Logs, and Log of Communications," MSI.

190 "Well, let's get her over with.": *Wisconsin State Journal*, July 29, 1945, 4.

190 It's as if they have stepped into the *Flying Dutchman*: Gallery, *Twenty Million Tons*, 234.

191 require a steady source of seawater to flush the toilets, cool the engines, and feed the freshwater distiller: *Hunt and Kill*, 154.

191 "Wayne, here goes nothin'," Lukosius tells him: Lukosius, oral history, MSI, 14.

192 The whaleboat is getting so crammed and Beaver so harried he wonders if he should jettison unimportant items: Pickels war diary, June 4, 1944.

193 Wayne Pickels passes the mess hall: Pickels war diary, June 4, 1944.

194 "Water is coming in the hatch," Hohne hollers: *Wisconsin State Journal*, July 29, 1945, 4.

195 Plus a Belgian Mauser revolver and a leather officer's coat: Pickels war diary, June 4, 1944.

Chapter 26: "Who's in Charge?"

197 Hans Goebeler is on the life raft swaying in the ocean, grateful to be alive: Goebeler, 239.

198 "Put it in your mouth so they won't find it," a German tells him: Werner Reh, Oral History Project, MSI, June 2, 19999, 10.

199 "The captain is shot up and can't talk," Lieutenant Commander Dudley Knox radios: Action Reports, MSI.

199 "Commander Trosino, report to the captain on the bridge immediately": Trosino oral history, MSI.

201 "Yeah," Mama Trosino answered. "He's upstairs asleep.": Ibid.

201 "Lieutenant Hampton report to Captain Gallery on the bridge": Deward "Hap" Hampton, Oran History Project, March 3, 1999, 12.

202 he's claustrophobic: Ibid.

202 I'm from a barely there farm in Alabama: Ibid.

202 "Grab the gear!" Trosino calls out: Trosino, oral history, MSI.

203 "doggone weak": Ibid.

204 "Come on, bud, get up," Trosino snaps. "Out of here.": Ibid.

204 "sixty million fleas": Ibid.

205 "a great big, huge tube of nothing but values and machinery": Ibid.

205 "Well, I'm appreciative. Misery loves company.": Ibid.

206 Mocarski's deep blue eyes: *Wisconsin State Journal*, July 29, 1945, 4.

207 "I don't know, Mr. David. I just ain't talking.": Ibid.

Chapter 27: Request Immediate Assistance

209 "We backed away from the sub now. Her planes made a big hole in the side.": Action Reports, MSI.

210 he's worried some dimwit will accidentally push the wrong button: Gallery, Ibid.

210 "taut as a banjo string," Gallery, *Twenty Millions Tons*, 286.

211 "Get 'em off the ship," Trosino growls: Trosino, oral history, MSI.

211 crawling through oily water: Gallery, *Saturday Evening Post*, "We Captured a German Sub," August 4, 1945, Vol. 218, No. 5; 9.

212 He calls Trosino by his first name: Ibid.

212 Lieutenant David may have captured the sub, but it's Trosino who saved it: Gallery, *Clear the Decks!*, 209.

212 He's got a bottle of Canadian Club in hand to celebrate: Gordon Hohne, Oral History Project, date unknown, MSI.

212 Flags Hohne is having a delayed reaction to the day's events: Ibid.

213 All the cooks can drum up is a bologna sandwich: Pickels diary, June 4, 1944.

213 REQUEST IMMEDIATE ASSISTANCE TO TOW CAPTURED SUBMARINE U-505: MSI.

215 "THIS IS FOR YOUR FILES REGARDING: Gallery, *Twenty Millions Tons*, 238.

Chapter 28: Junior

216 On the *Jenks*, Paul Meyer, second in command of *U-505*, is undergoing inter-rogation: Medical Department to Representatives, Command Officer USS *Jenks*, June 6, 1944, MSI.

216 They have better luck squeezing *U-505* prisoner Heinrich Brall: Action Reports, USS *Jenks* DE665, FILE DE665 TE/A 16-3, Lt. Cmd. J.F. Way, June 7, 1944, MSI.

217 To relax and make them feel at home, a Viennese waltz is played: Joseph Villanella, Oral History Project, March 5, 1999, MSI, 3.

218 They're strip-searched one more time, including a cavity probe: Springer, Oral History, MSI, 22.

218 the rest of the POWs are locked in a caged pen located below the flight deck: Goebeler, 245.

218 Who does he think he is—Wyatt Earp?: Gallery, *Clear the Decks!*, 210. (In Gallery's book, he actually makes the comparison to Hopalong Cassidy.)

219 they are not permitted to shave or get a haircut: Schiller, oral history, MSI, 7.

219 Lieutenant Jack Dumford finds himself in the intimidating presence of the great man: Dumford, oral history, MSI.

220 "You shouldn't have had pork chops": Dumford, Ibid, 45.

221 Gallery sits there at his table slurping tomato soup: Ibid., 27.

222 He isn't certain he has "any business being there.": Gallery, *Twenty Million Tons*, 239.

223 "You can't live forever": Ibid.

224 "Wait a minute," Trosino says: Trosino, oral history, 30.

Chapter 29: "I Will Be Punished"

225 "Look here, boy. You are in a very bad situation with your leg," Gallery says: Harald Lange interview, *Kristall* magazine, Hamburg, 1956.

226 I ordered the men around me to give three cheers for our sinking boat: Lange, Statement of Commanding Officer of *U-505*, MSI.

226 They are Morse code dots and dashes, scratched into the prison bars—left behind by the crew of *U-505*'s sister boat: Goebeler, 244.

227 "I'm a soldier of the German Reich": Related by Cmdr. Earl Trosino, oral history, MSI, 34.

227 Leon "Ski" Bednarczyk was born in New Haven, Connecticut, in 1918: back-ground information based on interview with Leon Bednarczyk's son, Ed, August 24, 1922.

228 "Ski, we're losing the sub": Ibid.

228 "Does anybody here speak Polish?" he asks: Ibid.

230 "Relax," Trosino tells him: Trosino, MSI, 35.

231 "if we don't keep that covered and cool it will explode": Ibid., 44.

232 "Would you be willing to spend the night on board?": Trosino, MSI, 46.

233 Felix promises to remain for the "rest of his life and never ask for pay": Ibid., 43.

233 Felix is as "healthy as an ox.": Goebeler, 245. Goebeler remembered the "cause of death" to be tuberculosis. Other accounts say Gallery attributed it to a "stomach ailment."

Chapter 30: USS *Nemo*

235 Towing *U-505* reminds him of a dachshund being dragged by an elephant: suggested by author of *Kristall* magazine article.

235 "Take charge of the first thing you see," it reads: Horace Mann Jr., "My Memories of the Navy," Navy Memorial Log, date unknown.

236 *U-505* is back "in the bag.": Gallery report to Dept. of Navy, date unknown, MSI.

236 but to Gallery she is the most beautiful ship in the Navy: Gallery, *Clear the Decks!*, 209.

236 "How'd you make out with Commander Rucker?": Trosino, oral history, MSI, 39.

237 IN VIEW OF THE IMPORTANCE AT THIS TIME OF PREVENTING: dispatch COMINCH to Admiralty, MSI.

238 King even threatens to court-martial Gallery: Herb Gilliland, Coda: The Last Secret of the *U-505*, published by MSI.

238 EMPHASIZE TO ALL CONCERNED NEED FOR ABSOLUTE SECRECY REGARDING CAPTURE.": Dispatch King TO CINCLANT, MSI.

239 "Tomorrow morning I want you to take down the flags": Don Carter, Oral History Project, March 3, 1999, MSI.

240 "We want to know what happened to the German flag": Ibid.

242 "There is no use whatsoever having a souvenir unless you can show it around": Gallery, *Clear the Decks!*: 212.

242 the "damnedest collection of stuff": Gallery, *Twenty Million Tons*, 244.

243 on the bridge, at midnight, when a notion strikes him: Hohne, oral history, MSI.

243 "I can't tell you where I am," the letter goes, "but I'm on KP": Ritzdorf, oral history, MSI, 19.

Chapter 31: Where Are My Men?

247 "rather decent sort of chap": Gallery, *Saturday Evening Post*, 74.

247 "They can't do that!" declares the newspaper editor. "This is not American territory.": Andrew Birmingham, Bermuda Military Rarities Revisited, Bermuda Historical Society, 2012, 67.

248 "Tell me," Bishop says, "what's that rusty submarine doing out there?": Jim Bishop, "Big War Story Hushed Up," *San Francisco Examiner*, July 31, 1958.

Author's Note: This is the same Jim Bishop who wrote the bestsellers *The Day Lincoln Was Shot* and *The Day Kennedy Was Shot*.

248 SUBJECT NEMO IS BEING DISCUSSED IN NORFOLK AREA BY THOSE WHO DO NOT NEED TO KNOW: Admiral F.S. Low to CINCLANT: June 15, 1944, MSI.

249 "absolute security concerning NEMO" must be maintained: Memo from Low to Assistant Chief of Staff," June 18, 1944. SUBJECT: ITINERARY FOR BERMUDA TRIP.

249 "Hallelujah!" somebody shouts: Dumford, oral history, MSI, 28.

250 When he is finished, Dumford is told he is "free to go": Ibid., 29.

250 That frees up thirteen thousand hours of precious decoding computer time: Navy Dept. memo, Office of the Chief of Naval Operations, June 13, 1944, to OP-20, Cmdr. J.N. Wenger, MSI.

251 In spite of several attempts to establish communications and a certain waiting period, [*U-505*] did not respond: Letter to Heinrich Goebeler from Kreigsmarines, MSI.

252 "If ever an MIA notice should come, then you can count on it that I am no longer alive.": Springer, oral history, MSI, 7.

252 "Please, Frau Lange, tell me honestly—is your husband a daredevil wanting to win the Knight's Cross?": *Kristall* magazine.

253 There are maids and a cook named Queenie, a grass tennis court, squash court, and boathouse. Dinner is formal at Paget Hall: Paget Humphries Dackman interview, August 14, 2022.

254 One day she is summoned to Admiral Frank Braisted's office: "Recollections of Mrs. Shirley Humphries," Bermuda Military Rarities Revisited, 73.

254 He tells Shirley that he's worried she might think he is dead and remarry: Ibid.

Chapter 32: Medal of Honor

256 Werner Reh calls it schlaffaria—a wonderland, paradise: Reh, oral history, MSI, 15.

256 "I'm not allowed to say anything," Schiller tells him: Schiller, oral history, MSI, 9–10.

257 Hans Goebeler points the finger at Chief Engineer Josef Hauser: Goebeler, 247.

257 "highest security," goes King's memo: Waller, 5+.

257 Petty Officer Karl Springer rounds one corner and comes upon two nauseous leathernecks: Springer, oral history, MSI.

258 The Germans are deloused, interrogated, and issued black PW clothes. But they're not staying very long: Derek Waller, "What Happened to U-505's Crew

After Their Capture" (*Argonauta*, Winter 2020) 5+. Author's note: Waller is a retired air commodore with the Royal Air Force who has done extensive research on German U-boats.

258 Gene Moore, a hot prospect for the Brooklyn Dodgers baseball team: Gary W. Moore, *Playing with the Enemy: A Baseball Prodigy, World War II, and the Long Journey Home* (New York: Penguin Books, 2008).

259 UPON ARRIVAL AT CAMP RUSTON, LA . . . THIS GROUP SHOULD BE KEPT ENTIRELY SEPARATE (*Argonauta*, Winter 2020) 5+.

259 That means no mingling with the two thousand German prisoners: memo from Major Gen. Clayton Bissell to Provost Marshal General, War Dept., Washington, DC, January 25, 1945, MSI.

260 "We've done something nobody else has ever done.": Interview with Ed Bednarczyk (Leon's son).

261 King presents *Roosevelt Kampf* to none other than President Roosevelt: Thomas B. Buell, *Master of Seapower: A biography of Fleet Admiral Ernest J. King* (Naval Institute Press, Annapolis, Md., 1980), 197.

262 "Well, son, she turned out to be a pretty good ship after all": Gallery, *Clear the Decks!*, 218. (Author's note: Gallery wouldn't publish the expletive. He wrote that it was "uninhibited and unprintable.")

262 but he's "champing at the bit" to get back to the war: Gilliland & Shenk, 133.

262 "There is no greater demotion for a naval officer than to go from the bridge of a combatant ship: Gallery, *Eight Bells* (New York: W.W. Norton, 1965), 185.

Chapter 33: They're Alive!

265 When Werner Reh breaks his toe playing a game of soccer: Reh, oral history, MSI, 20.

265 The POWs are full of bluster about burrowing under the fence and making their way to South America: Schiller, oral history, MSI, 24.

266 they merit a nickname—"*stuka*"—after the Luftwaffe dive-bomber: Goebeler, 249.

266 claim that the site is "forbidden" to civilians: Waller, quoting the International Red Cross, 10.

267 An investigation is launched and the Kalbitz-Geuthner connection is ferreted out: Naval Prisoner of War Section, Postal Censorship Branch: Ibid., 11.

267 "How would I know? What a stupid question!": Birmingham, 67.

268 In Paris, sirens wail and cannons boom and the famed boulevards: *New York Times*, May 8–9, 1945.

270 Bode hoists a black flag, signaling submission: United Press, May 14, 1945.

270 "Your son and the crew are doing well," Carla Lange: *Kristall* magazine.

Chapter 34: Now It Can Be Told

272 "The boys did keep their mouths shut": Gallery, *Twenty Million Tons*, 244.

272 the "greatest [intelligence] windfall of the war": Gallery, *Saturday Evening Post*.

275 The cause of death is determined to be coronary occlusion: Certificate of Death, Commonwealth of Virginia, State File No. 18725.

275 Mack Pickels is let out on Christmas Eve and returns home to San Antonio, Texas: Orem, Wayne Macveigh Pickels, 62.

275 "Let's get married on June fourth," he says: story related by Douglas Pickels.

276 "It's either me or the Navy": Hohne family interviews.

276 Knispel harbors a little resentment: Hohne, oral history, MSI.

277 "Ski, you know I'm not going to be able to write about what you did. It's classified: conversation between Gallery and Leon Bednarczyk, related by Ed Bednarczyk.

278 "I was putting out fires at the school and she was putting out fires in me": Trosino, unpublished biography he wrote for *Power Magazine* (McGraw-Hill Publishing), May 17, 1957, courtesy: Trosino Family.

278 *Guadalcanal* will end her days sold, wouldn't you know it, to the Japanese. For scrap: Gallery, *Clear the Decks!*, 218. Author's Note: Rear Admiral Gallery made one last landing and takeoff from the ship, in a helicopter, off Guantanamo, Cuba.

Chapter 35: "Dear Mother!"

279 He's in his barracks at Camp McCain. Springer has taken up knitting: Springer, oral history, MSI, 19.

280 "I give you my word of honor as an American officer": Springer, oral history, MSI, 17.

281 AFTER TWO YEARS, I AM AGAIN ON THE COAST OF WESTERN EUROPE: Hauser letter, courtesy of MSI.

282 "revolted" at the thought of swearing an oath to an "enemy-imposed regime": Goebeler, 250.

282 "Dear Friend," the letter begins in broken English. "Now it is nearly two years ago that we parted at Norfolk, but I didn't forget going to *Guadalcanal* yet: Ewald Felix to Leon Bednarczyk, June 10, 1946, courtesy of MSI.

283 Gallery heads over to the State Department and meets with the head of the Visa Section: Trosino letter to Felix, March 4, 1948, courtesy of Trosino family.

283 "I know you will be pleased and see that I am trying to do all that is possible to get you here": Trosino letter to Felix, March 4, 1948.

283 "invaluable assistance . . . in the prosecution of the war effort": NARA, Washington DC i.g. 38 Office of Naval Intelligence, Special Activities Branch, cited by Waller, 189.

284 "A truly great man, and to this day I stand by him," Loewe says: Loewe letter to Gallery, September 29, 1955, letter addressed, "My Most Honoured Admiral." MSI. Author's note: Gallery's opinion of Karl Doenitz was complex. He called Doenitz's conviction at Nuremberg "an insult to our own submariners. When the United States got into the war . . . the Pacific operated the same as the Germans did, and sank 6,000,000 tons of ships. We torpedoed without warning. . . . That's the only way submarines can operate. War is a brutal business and no amount of wishful thinking by a bubbleheaded statesman can make it otherwise." Gallery, however, ignores other aspects of Doenitz's career, including his unrelenting support for Hitler and the Nazi Party, fueled by implacable anti-Semitism.

Epilogue: The Quiet Men

285 "evil in her looks and a problem to handle": Trosino, unpublished *Power* magazine manuscript.

286 "Are you kidding?" Trosino wrote back: *Hunt and Kill*, 192. This chapter was written by Keith Gill, formerly of the Museum of Science and Industry, Chicago.

287 On the evening of September 3, after the evening rush hour: Gallery, *Twenty Million Tons*, 256. See Gallery's book for details.

287 A choir sings "Eternal Father, Strong to Save," the Navy Hymn. A Marine honor guard fires a volley: Gill, *Hunt and Kill*, 215.

288 "Cap'n, where the hell is that Luger you made me turn in?": Gallery, 257.

288 Chester Mocarski never told his wife how he nearly lost his life: Gill, *Hunt and Kill*, 213.

289 "Guess we're going to Chicago," Hohne said: Pamela Genelli interview.

290 "Yesterday's enemies are today's friends," Pickels liked to say: Douglas Pickels interview.

290 Villanella had to chuckle. That was a name he could never forget: Villanella, oral history, MSI, 8.

292 Stanley "Stash" Wdowiak didn't want to talk about his war experiences either: Nick Mascia interview, July 19, 2022. Mascia is Wdowiak's son-in-law.

293 "We can't go up there!" Trosino called out: *Chicago Tribune*, June 14, 1989.

293 "As you may have heard," he wrote Loewe, "*U-505* was eventually captured: Gallery letter to Loewe, July 28, 1955, MSI.

294 "I would have been proud to have had you as a crew member of my ship.": *Chicago Daily News*, September 11, 1958.

295 he "couldn't put it down—and sat up till 5:00 a.m. this morning to finish it.": Gilliland and Shenk, Gallery to Herman Wouk, 162.

295 "A short lean man in a brown jacket and a bow tie, with large ears and a quiz-zical face.": Herman Wouk, "A Reminiscence," in *Admiral Dan: The Life and Wit of a Navy Original* by Gilliland and Shenk. Author's note: Gallery served as Wouk's unofficial naval consultant for his bestselling novels, *Winds of War* and *War and Remembrance.*

296 "This has been one of those days when things just didn't seem to gel": *Maryville Daily Forum,* June 1, 1964, 1.

296 "Nobody here seems to know the hero": *Maryville Daily Forum,* June 1, 1964, 1.

SELECTED BIBLIOGRAPHY

Bermingham, Andrew P. *Bermuda Military Rarities Revisited*. Hamilton, Bermuda: The Bermuda Historical Society, 2012.

Buell, Thomas B. *Master of Seapower: A Biography of Fleet Admiral Ernest J. King*. Annapolis, MD: Naval Institute Press, 1980.

Carey, Alan G. *Sighted Sub: The United States Navy's Air Campaign Against the U-Boat*. Havertown, PA: Casemate Publishers, 2019.

Gallery, Daniel V. *Clear the Decks!* New York: Paperback Library, 1967.

Gallery, Daniel V. *Eight Bells*. New York: Paperback Library, 1968.

Gallery, Daniel V. *Twenty Million Tons Under the Sea: The Daring Capture of the U-505*. Washington, DC: Regency Publishing, 1956.

Gilliland, C. Herbert, and Robert Shenk. *Admiral Dan Gallery: The Life and Wit of a Navy Original*. Annapolis, MD: Naval Institute Press, 1999.

Goebeler, Hans, and John Vanzo. *Steel Boat, Iron Hearts: A U-Boat Crewman's Life Aboard U-505*. El Dorado Hills, CA: Savas Beatie LLC, 2008.

Hardy, Clare. *SS Ceramic: The Untold Story*. Scotts Valley, CA: CreateSpace Independent Publishing Platform, 2012.

Kahn, David. *The Codebreakers: The Comprehensive History of Secret Communication from Ancient Times to the Internet*. New York: Scribner, 1967.

Kohnen, David. *Commanders Winn and Knowles: Winning the U-Boat War with Intelligence, 1939–1943*. Krakow, Poland: Enigma Press, 1999.

Kohnen, David. "Tombstone of Victory: Tracking U-505 from German Commercial Raider to American War Memorial 1944 to 1954." Phoenix: *Journal of America's Military Past*, Winter 2007.

Moore, Gary W. *Playing with the Enemy: A Baseball Prodigy, a World at War, and a Field of Broken Dreams*. New York: Penguin Books, 2006.

Mulligan, Timothy P. *Lone Wolf: The Life and Death of U-Boat Ace Werner Henke*. Westport, CT: Praeger Publishers, 1993.

Orem, Harry B., editor, Wayne Macveigh Pickels, BM 1/c, United States Navy, Silver Star Recipient. San Antonio, Texas: Privately Printed, 1999.

Savas, Theodore P., editor. *Hunt and Kill: U-505 and the U-boat War in the Atlantic*. El Dorado Hills, CA: Savas Beatie LLC, 2008.

Sebag-Montefiore, Hugh. *Enigma: The Battle for the Code*. Hoboken, NJ: John Wiley & Sons, 2000.

Sillcock, Ken. *Two Journeys into Peril: Wartime Letters of Ken and Ron Sillcock, 1940–1945*. Melbourne: Clouds of Magellan, 2009.

Smith-Hutton, Henri. *Reminisces of Captain Henri Smith-Hutton*. Annapolis, MD: US Naval Institute Press, 1976.

Tarrant, V. E. *The Last Year of the Kriegsmarine, May 1944–May 1945*. Annapolis, MD: Naval Institute Press, 1994.

Wise, James Jr., *Sole Survivors of the Sea*. Annapolis, MD: Naval Institute Press, 201.

INDEX

ABOUT THE AUTHOR

CHARLES LACHMAN is executive producer of the television news magazine show *Inside Edition*. Lachman is the author of *Footsteps in the Snow, The Last Lincolns: The Rise & Fall of a Great American Family, A Secret Life*, and the novel *In the Name of the Law*.